ENCYCLOPAEDIA OF EMINENT THINKERS

VOLUME-XI: **DADABHAI NAOROJI**

Volumes by the same author

The Political Thought of ***Dadabhai Naoroji***
The Political Thought of ***Tej Bahadur Sapru***
The Political Thought of ***C. Rajagopalachari***
The Political Thought of ***Subhas Chandra Bose***
The Political Thought of ***Bhagat Singh***
The Political Thought of ***Annie Besant***
The Political Thought of ***C.R. Das***
The Political Thought of ***Lalalajpat Rai***
The Political Thought of ***Sir Pherozeshah Mehta***
The Political Thought of ***Mahadev Govind Ranade***

About the Author

Dr. (Miss) Ashu Pasricha is a faculty in Gandhian studies at Panjab University, Chandigarh. She has many books to her credit including *Gandhian Approach to Integrated Rural Development; Public Administration: Discipline and Dimensions; Peace Studies: Discipline and Dimensions; WTO, Self-reliance and Globalisation; and Gandhi's Concept of State and Power.* Also, she has published a number of research papers in various Indian and foreign journals of repute.

ENCYCLOPAEDIA OF EMINENT THINKERS

VOLUME ELEVEN

POLITICAL THOUGHT OF DADABHAI NAOROJI

Dr. Ashu Pasricha
Department of Gandhian Studies
Panjab University, Chandigarh

CONCEPT PUBLISHING COMPANY, NEW DELHI-110059

Cataloging in Publication Data—DK
Courtesy: D.K. Agencies (P) Ltd. <docinfo@dkagencies.com>

Encyclopaedia of eminent thinkers / K.S. Bharathi.
v. 11-15 ; cm. — (Gandhian studies and peace research series ; 11)
Vols. 11, 14-15 by Ashu Pasricha; v. 12-13 by Jai Narain Sharma.
Includes bibliographical references.
Includes indexes.
ISBN 13: 9788170226840 (set)
ISBN 10: 8170226848 (set)

1. Statesmen—India. 2. Political science—India. 3. India—Politics and government. I. Bharathi, K. S. (Kuduva Swamy), 1956- II. Pasricha, Ashu, 1967- III. Sharma, Jai Narain, 1951- IV. Series: Gandhian studies and peace research series ; 11.

DDC 954.035 22

ISBN -13: 978-81-7022-684-0 (Series)
ISBN -10: 81-7022-684-8 (Series)
ISBN -13: 978-81-8069-491-2 (Vol. XI)
ISBN -10: 81-8069-491-7 (Vol. XI)

First Published 2008

Published and Printed by
Ashok Kumar Mittal
Concept Publishing Company
A/15-16, Commercial Block, Mohan Garden
New Delhi-110059 (India)
Phones: 25351460, 25351794
Fax: 091-11-25357103
E-mail: publishing@conceptpub.com

DEDICATED
TO
MY PARENTS
SHRIMATI SUSHILA PASRICHA
and
SHRI B.R. PASRICHA

PREFACE

The early history of the Indian National Movement is studied with a few precious jewels of Indian liberals who had toiled and worked hard with great vigour and devotion for emancipation of India, and contributed considerably to India's freedom. Dadabhai Naoroji the Grand Old Man as he was popularly known, was amongst the first early Indian patriarchs of Indian nationalism. As a great nationalist leader he advocated the Indian interests and served India with 'undeviating purpose with complete selflessness and vitality of purpose'. He associated himself with the national organisation since its inception, and made efforts to further the Indian national cause till the sunset of his eventful life.

A liberal and moderate, he was the intellectual child of European liberalism. His great merit lay in being the interpreter of European liberal ideas to the people of India. As a liberal leader he believed in the British sense of justice and fair play. He admired the British culture, and believed that a link with Britain signified India's welfare. He had firm faith in the constitutional means of agitation which could fulfil the object of political advancement of India under British rule. By virtue of his spirited leadership he had the honour of chairing the Congress thrice in 1886, 1893 and 1906; and throughout his public life he held aloft the banner of the national organisation.

It is difficult to recall any Indian intellectual using a fairly empirical methodology to such a good macro purpose as Dadabhai Naoroji had done. Since he was well familiar with free trade doctrine of British classical economists and had conviction that the doctrine was in the interest of those entering into international trade; further since he knew that both sides in such trade gained in utility, he naturally devoted his attention to finding how India and Britain gained from exports and imports to each other. He applied his studiousness to collect and work out statistics and was rewarded by the conclusion that the trade relationship between the two was not to their mutual advantage. In fact, he found, that what to say of any advantage, the relationship was one sided and indicated unilateral transfer of surpluses from India to England for a fairly large spread of time. This gave him some clue to India's abysmal poverty at that time. The continuing drain of our resources through disadvantageous trade checkmated higher investment and capital formation and prevented any positive impact on the impoverishment of the Indian people.

Dadabhai Naoroji used empirical data to prove that there was drain of wealth from India even in its external trade. He also discovered that there existed internal drain in India, in the sense that people in India had to part with higher proportion of their capacity to pay to the government treasury than paid by people in England. Much of this money went to reimburse the fabulous salary of British bureaucracy. Another major contribution of Dadabhai Naoroji was to focus on India's national and per capita income and India's poverty through linking foreign trade with drain.

Dadabhai's ideas seem to contradict the prevailing paradigm that globalisation through increasing both the volume of international trade and the share of both the developing and developed countries in it, would result in increased welfare of all the trading partners.

Personal acknowledgements are no easy matters where so many have been kind. My special thanks are due to Professor R.N. Pal, Professor N. Radhakrishnan, Dr. K.S. Bharathi, Dr. A.D. Mishra, Professor M.L. Sharma, Professor Jai Narain Sharma, Dr. Meena Dutta, Dr. Seema Pasricha, Dr. Rashmi Sharma, Mr. Ajay Pasricha (Advocate) and Master Hemant Malhotra. Beyond that I can make only an all embracing salute to the galaxy of the scholars whose works I have consulted.

I am also grateful to Shri Ashok Kumar Mittal, my publisher, for his consistent zeal in the printing and publication of this volume.

Chandigarh **ASHU PASRICHA**

CONTENTS

Encyclopaedia of Eminent Thinkers Series

Vol. I. The Political Thought of **Mahatma Gandhi**/ *K. S. Bharti*
Vol. II. The Political Thought of **Nehru**/ *K. S. Bharti*
Vol. III. The Political Thought of **Rabindra Nath Tagore**/ *K. S. Bharti*
Vol. IV. The Political Thought of **Radhakrishnan**/ *K. S. Bharti*
Vol. V. The Political Thought of **Vinoba**/ *K. S. Bharti*
Vol. VI. The Political Thought of **Aurobindo**/ *K.S. Bharti*
Vol. VII. The Political Thought of **Jayaprakash Narayan**/ *K. S. Bharti*
Vol. VIII. The Political Thought of **Vivekananda**/ *K.S. Bharti*
Vol. IX. The Political Thought of **Ambedkar**/ *K. S. Bharti*
Vol. X. The Political Thought of **M.N. Roy**/ *K. S. Bharti*
Vol. XI. The Political Thought of **Dadabhai Naoroji**/ *Ashu Pasricha*
Vol. XII. The Political Thought of **Veer Savarkar**/ *Jai Narain Sharma*
Vol. XIII. The Political Thought of **M.A. Jinnah**/ *Jai Narain Sharma*
Vol. XIV. The Political Thought of **Tej Bahadur Sapru**/ *Ashu Pasricha*
Vol. XV. The Political Thought of **C. Rajagopalachari**/ *Ashu Pasricha*
Vol. XVI. The Political Thought of **Subhas Chandra Bose**/ *Ashu Pasricha*
Vol. XVII. The Political Thought of **Mohammad Iqbal**/ *Jai Narain Sharma*
Vol. XVIII. The Political Thought of **K.B. Hedgewar**/ *Jai Narain Sharma*
Vol. XIX. The Political Thought of **Bhagat Singh**/ *Ashu Pasricha*
Vol. XX. The Political Thought of **M.S. Golwalkar**/ *Jai Narain Sharma*

1

LIFE AND TIMES

The early history of the Indian National Movement is studded with a few precious jewels of Indian liberals who had toiled and worked hard with great vigour and devotion for emancipation of India, and contributed considerably to India's freedom. Dadabhai Naoroji the Grand Old Man as he was popularly known, was amongst the first early Indian patriarchs of Indian nationalism. As a great nationalist leader he advocated the Indian interests and served India with 'undeviating purpose with complete selflessness and vitality of purpose'. He associated himself with the national organisation since its inception and made efforts to further the Indian national cause till the sunset of his eventful life.

Dadabahi's long life spanned three quarters of the nineteenth century and seventeen years of the twentieth. It falls into four or five periods: The first, the period of childhood and youth, from 1825 to 1855; the second, from 1855, when Dadabhai first went to England, upto 1881 when he left England with the intention, as he then thought, of never returning again; the third, from 1881 to 1896, when he did return to England with the intention of trying to enter Parliament; the fourth, from 1886 to 1907, the year when he finally returned to India; and the fifth, the ten years of retirement from 1907 upto his death in 1917.

For the first period (1825-1855) we are fortunate in-having Dadabhai's own account of the influences of his childhood and youth that shaped his life and character.[1] He tells us how the reading of the *Shahnama* in Gujarati, and recounting the heroic deeds of long past Persian ancestors, was the inspiration of his early childhood (just as many a Hindu lads have been enthused by the stories of the

Ramayana and the *Mahabharat*); how he was lucky enough to obtain education, vernacular and English, in the then recently established Native Education Society's school, without having to pay any fees (which his mother could not have been afforded); how he distinguished himself among his fellow students, in games as well as in his studies, and became the regular 'show-boy' of his class; how, when he came to years of discretion, the idea occurred to him that as he had received his education at the expense of the people, he must make a return to them with the best that he had in him, and how this idea was reinforced by his reading of certain books like Clarkson's on the Slave Trade and the life of Howard the philanthropist; and how, during the six or seven years after he had finished his education, while he held a teaching post in his own *alma mater*, he and some of his young friends, with nothing in the pocket but with enthusiasm in the heart, encouraged by some of their elders like Jugonnath Sankarsett, Jimsetjee Jeejeebhoy, Professor Patton, and Sir Erskine Perry, threw themselves heart and soul into a number of reform movements—movements for the education of girls, for diffusion of useful knowledge in the vernaculars, for religious reform, for education of the public through the Press, for political reform, etc. The Students' Literary and Scientific Society was founded in 1848 for the education of girls, the Dnyana Prasarak Mandli for diffusion of useful knowledge in the same year, the Rahnumai Mazdayasnan and the *Rast Goftar* for Parsee religious reform and for education of the public, respectively, in 1851, and the Bombay Association for voicing of the political needs of the people, in 1852.[2] Dadabhai was Treasurer and Vice President of the Students' Literary and Scientific Society, President of the Gujarati Dnyana Prasarak, Secretary of Rahnumai Mazdayasnan between 1851 and 1853; editor of the *Rast Goftar* for 2 years; and a member of the Committee of the Bombay Association. He read several papers before the Students' Literary and Scientific Society and the Dnyana Prasarak.[3]

The second period, from 1855 to 1881, is a crucial one in Dadabhai's life. It was during this period that his ideas on the economic and political problems of India became fixed. It was also a period of many sided activities, on his part.

During these 25 years, Dadabhai moved several times between England and India. For the first three years, he represented his firm (Cama & Co.) at Liverpool. He left the firm in 1858, and returned to India, but went to England again the next year, and established a firm of his own, by name Dadabhai Naoroji & Co. In September 1863 he was back home once more, and stayed for a year and a half, returning to England in April 1865. He was again in Bombay for a few months in 1869. The East India Association, in the establishment of which Dadabhai had taken a leading part, had been founded in London in December 1866, and during his brief sojourn, Dadabhai established a Branch of it in Bombay. It was at this time also that he received an address and a purse from the citizens of Bombay for his public services.

In 1871 Dadabhai came once more to India to canvass support for the East India Association. He returned to England in 1872, only to be urgently called back by the Maharaja of Baroda towards the end of 1873, with the offer of the Diwanship of his State.[4] Dadabhai served in that capacity for just one year 1874. The period between January 1875 and August 1876 was spent in Bombay. Dadabhai was elected member of the Municipal Corporation and the Town Council in July 1875, but he resigned that office in September 1876, and sailed once more for England. He spent the next five years in England, and returned to India in 1881, intending to stay there for good.

A visitor from India to England towards the middle of the 19th century could not fail to be impressed with the greatness of England, her wealth, her power, and her prestige. England was at this time the greatest power in Europe. Alone of the powers of Europe, she had not bent before Napoleon. She was now mistress of the seas, and of an empire on which the Sun never set. Her advanced industrial technique had made her rich beyond measure. Her political institutions were ahead of other nations, and while elsewhere people were trying, through rebellions and revolutions, to overthrow autocratic regimes, she stood as an exemplar to the world of the reconciliation of liberty with order, and was providing an asylum for popular leaders from other countries fleeing from the wrath of their rulers. That she had, at a sacrifice of £ 20,000,000 freed the slaves in her colonies, and that she was granting self-government to her colonies, was proof that the nation was moved by moral considerations, and that the Englishman's love of liberty was not confined to himself. In India, too, the English had not only established law and order, but they were educating the people, with clear recognition that 'having become instructed in European knowledge,' the people 'may, in some future age demand European institutions'. Is it then to be wondered at that Dadabhai should pin his faith to the noble and freedom loving British nation in his fight for justice to India?[4]

It is very likely that it was his experience of England that burnt the fact of Indian poverty into Dadabhai's soul, and set him thinking of its causes and its cure. The six or seven years previous to 1855 were years of hectic reforming activity but that activity was confined to social, educational and religious reform (with a very mild beginning of political reform), and the problem of poverty was not so much as thought of. It is true that two writers, Ramkrishna Vishwanath, the famous 'Lokahitwadi', and Gopalrao Hari Deshmukh had already, in 1843 and 1849 respectively, written about the poverty of the people, its causes, and the remedies for it, but they wrote in Marathi and their writings did not seem to have come to Dadabhai's notice. In a few years after 1855, however, Indian poverty becomes the one problem which absorbs all Dadabhai's attention. Its cause is traced to the drain of India's wealth, itself the natural and inevitable result of alien rule and the employment of foreign agency in the administration of the country; and increasingly with the passage of years relief is sought mainly through the substitution of native for foreign agency. Great and growing poverty, the drain, and simultaneous examinations for the Services hereafter become the main, almost the exclusive themes of Dadabhai's thought and propaganda.

Even before the establishment of the East India Association, Dadabhai was active in promoting, whenever possible, the interests of India and the Indians. His correspondence with Secretary of State on the question of the age limit for the Civil Service has been already referred to. He became a member of several learned Societies, and before one of them, the Literary and Philosophical

Society of Liverpool, he read papers in 1861 on the 'Manners and Customs of the Parsees' and on 'Parsee Religion' in 1865 he took the lead in making a representation to the Secretary of State on the question of the marks to be allotted to Sanskrit and Arabic in the Civil Service examination; in 1866 he read a paper before the Ethnological Society. London on "European and Asiatic Races", controverting the views enunciated by John Crawfurd, the then President of the society, about the superiority of European over the Asiatic races.[5] This was a spirited reply marked by much learning and cogent reasoning.

After the establishment of the East India Association, for the independent and disinterested advocacy and promotion by all legitimate means of the public interests and welfare of the inhabitants of India generally", Dadabhai found a ready forum for discussion of matters of interest to India. He read several papers before it, and took part in discussion on papers read by others. The first paper he read was on 'England's Duties to India' (May 2, 1867) in which he made mention of the continued loss of wealth to the country, and the great necessity of developing its resources by large investments. He pleaded also for greater participation by the Natives in the affairs of their own country, suggesting as mentioned above, the holding of simultaneous examinations, atleast for a part of the appointments.[6] In a paper on the "Wants and Means of India" read on 27th July 1870, he said that India's per capita income was only 40 Shillings per annum, and again he looked for salvation to large irrigation works, railways and roads, and other public works.[7] He was willing to accept that not all that India pays for the services of Englishmen is loss. This is what he says in a discussion on January 31, 1868 at the East India Association: "With respect to the economic results of the political relations between England and India .... if England goes to India to serve it, it must receive its wages. India is benefited by that service, though, as the English, their is a foreign rule, their wages go out of the country, while in the case of a native government, the wages would remain in the country, and be a part of the wealth of the country. India ought not to grudge these wages. The order and laws introduced into India by the Introduction of British rule, the changes which that rule has effected in its political, social, moral and intellectual condition are a sufficient benefit, and more than a sufficient benefit, for what it has to pay.[8] Similarly, in the paper on "The Commerce of India", read on 15th February 1871, after putting the question, "Why should India have to remit £ 12,000,000 a year to this country?", Dadabhai goes on to say, "This, to a certain extent, is inevitable. If India is to be regenerated by England; India must make up its mind to pay the price. The only thing I have to say is that England on its part also should act just towards India.... No unreasonable burdens should be unopposed on India because it is at your mercy; and the revenues of India should be administered with economy, wisdom and the sense of responsibility of a great trust."[9] In his evidence submitted to the Select Committee on East India Finance, 1871, Dadabhai contrasts the respective tax-paying capacity of England and India by saying that whatever revenue is raised in England remains in the country and so the national capital, upon which the production of a country depends, does not suffer diminution; while, on account of India's being subject to foreign rule, out of the £ 50,000,000 of revenue raised every year some £ 12,000,000 or more, are carried clearly away to England, and the national capital, or, in other words, its capability of productions continuously diminished year after year."[10]

The concessions which Dadabhai was prepared to make in these early years, that India must be prepared to pay the price for British rule, that the benefits of that rule outweighed what India had to pay for them etc., were ignored by him in his later writings and the disadvantages connected with the annual remittances to England, viz. that they took away so much from the potential productive capacity of the country alone loomed large before his eye, so that the whole of the 'Home charges' came to be characterized as a 'drain' (a word which seems to have been first used by Dadabhai in his paper on the 'Commerce of India'). So also, the importance of public works as an ameliorative factor was underplayed in later writings, and all importance was given to the substitution of foreign by Native agency in administration. Thus Poverty-Drain Simultaneously Examinations—became the burden of his propaganda for the rest of his life.

In 1876 Dadabhai was in India. He read an elaborate paper before the Bombay Branch of the East India Association on the Poverty of India. It must have been the result of many years of labour spent on the collection of statistics. It was printed as a pamphlet and distributed widely. It drew the attention of H.M. Hyndman, the Socialist leader, whose correspondence with Dadabhai dates from about this time.

In 1880 Dadabhai published another pamphlet under the title 'Condition of India', containing his correspondence with the Secretary of State on the subject of Indian poverty. In 1881 he submitted a Memorandum on some statements in the Report of the Indian Famine Commission 1880, in which he pressed the same view as in the earlier pamphlets, viz. "that the policy of English rule as it is (not as it can and should be), is an everlasting, unceasing and everyday increasing foreign invasion, utterly, though gradually destroying the country."[11]

When one considers the phenomenal activity of Dadabhai during this period and also the volume of his extant correspondence, one is disappointed to see so little of the latter bearing on this period. There is some correspondence early in the period with K.R. Cama on the subject of some promising young man to be brought to England to compete for the Civil Service, and on the conduct of the *Rast Goftar*, some with the Secretary of State on the lowering of the age limit for the Civil Service, on encouragement to female education; and on the expenses of the Indian army, sent on the Abyssinian expedition; some with Sir Erskine Perry mainly on the question of the Services; some bearing on his administration in Baroda,[12] and then for the last two or three years of the period, a few letters from Hyndman, a fair amount of correspondence with his family (this correspondence begins in 1879, when he sent his son Ardeshir to Bombay for his medical education), and a fair amount bearing on business. But there is no correspondence which might afford us a peep into the working of his mind on the questions on which his attitude became so fixed during this period. Did he not correspond at all with John Dickinson?[13] Evans Bell tells us in his Memoir of John Dickinson that "hardly any Indian nobleman or gentleman who has risen to distinction in any capacity, or whose name has become known in England within the last twenty years, has not at some time been on the file of Mr. Dickinson's correspondents".[14]

Dadabhai returned to India in April 1881, after closing his business in England (in which latterly he had not been doing too well). For the next five years he carried on some little business from Bombay. But these five years were also years of much public activity on his part. In 1882 he submitted a statement of his views to the Education Commission. In 1883 he started a few ventures —publication of a journal, under the title "The Voice of India", which gave extracts from important Indian journals on the leading topics of the day, the object being to enlighten the rulers about the real feelings of the people. The venture did not meet with the desired success. Dadabhai looked for support from wealthy well-wishers. He appealed to them to subscribe for as many copies as possible, so that the extra copies might be distributed gratis in England among influential men. "I have no doubt", he said, "the time will come when the Native public at large will understand and help in such efforts; but in the meantime, this duty must be performed by those who understand it. Much self-sacrifice in a variety of ways will have to be made by many a Native reformer before India rises among nations and before all her important reforms are achieved."[15]

Towards the end of 1883, Dadabhai's health broke down and he had to take rest for about six months, first at a place near Bassein, and then at Tithal near Bulsar. He had before this been elected to the Bombay Municipal Corporation. Owing to ill-health, he contemplated resignation, but was persuaded to continue as member by some friends. By June next year he had returned to Bombay, having improved somewhat in health, but still needing rest. That rest, however, was denied to him. The cause was the Ilbert Bill controversy. That Bill intended to abolish judicial disqualifications based on race distinctions provoked fierce opposition from Europeans all over the country. The Viceroy, Lord Ripon was abused and insulted; protest meetings were held and there was even talk of overpowering the sentries at Government House, putting the Viceroy on board a steamer and deporting him to England![16] At last the Government bent before the storm, and adopted a compromise, which, while doing away with the judicial inequality between Indian and European, and so in a way maintaining the principle of the Bill, introduced another inequality, viz. between Indian and European accused, by giving to a European accused the right of claiming a jury, half of whom at least must be Europeans. The compromise naturally led to intense disappointment among Indians, and for a time there was talk of a counter-agitation, especially in Calcutta. But the Bombay leaders took a more moderate stand point. They saw that if the result of the Ilbert Bill controversy was not to their liking, this was not Ripon's fault, and they gave him full credit for his good intentions. They decided, as Ripon was leaving India at the end of 1884, that he should be accorded a farewell such as had never before been accorded to a departing Viceroy. In spite of his weak health, Dadabhai threw himself heart and soul into the movement for a grand farewell and for a Memorial to Lord Ripon. Nothing would please the opponents—of Lord Ripon's policy so much, said Dadabhai, if, through our disappointment at the result of the Ilbert Bill agitation, we allow him to depart with a cold farewell. We must show that we understand and appreciate the spirit of his policy, that we know a true friend when we meet him, and are sensible of any good that is done to us.[17]

It was with a similar object that Dadabhai wrote to K.N. Kabraji, Editor of the *Rast Goftar*, from Tithal, suggesting that he might say a discreet word or two in his paper about the suitability of such men as Sir George Birdwood or Sir William Wedderburn as Governor of Bombay. It was necessary that Indians should make their voice heard in the appointment of their Governor, and say what qualities they expect in their Governors.[18]

Early in the next year 1885 came the "foundation of the Bombay Presidency Association. Since the Ilbert Bill controversy, Indian opinion was being estranged from the East India Association. It was felt that Association, which was largely supported by Indian money, was not serving Indian interests, and that it was necessary to establish another organization which would interpret Indian opinion better. The Bombay Presidency Association was accordingly inaugurated on 1st January 1885, with Sir Jamsetjee Jeejeebhoy as President, and Dadabhai as one of the Vice Presidents.

In view of the Acts passed in England in 1884 and 1885, extending the franchise, and redistributing the seats, a fresh general election was expected to take place at an early date, and the Presidency Association decided to send three delegates to England to educate the electorate about the necessity of reforms in India. Dadabhai exerted himself to obtain funds for the expenses of these delegates.[19]

September 1885, Dadabhai was nominated on the Bombay Legislative Council by Lord Reay, the Governor. His term as Councillor was very brief, as he left India for England in March 1896. But we can be sure that as a member of the Legislative Council, he would have conducted himself independently, without being unnecessarily offensive. That is his own view of how he would have conducted himself.

Later in 1885, Dadabhai along with Wedderburn, Hume, Mehta and Telang, was busy making preparations for holding the first National Congress in Bombay.[20]

This (1886-1907) period of 21 years was the culmination of Dadabhai's career. It was the time when his fame spread not only throughout the length and breath of India, but also to other countries of the world. To the people of India he seemed Patriotism incarnate, an embodiment of the nation's aspirations. He was called the "Grand Old Man" of the country. Such was the reverence universally felt for him that when in 1906 the question of the Congress Presidential election seemed about to lead to rupture among Congressmen, the proposal to elect Dadabhai was enough to heal it for the time being. A letter addressed to Dadabhai by Lokamanya Tilak in that year is subscribed, "Yours obediently, B. G. Tilak".

Dadabhai's visit to India in December 1906 to preside over the Calcutta Congress was his third visit during this period. Twice before he had made similar journeys in 1886 and 1893, and on both occasions he had presided over the Congress session of the year, at Calcutta and Lahore respectively. In 1886, after the Calcutta session, he spent a few days visiting some places in North India, but was hastily recalled by Malabari from Lucknow because Lord Rosebery, who was then in India, and was soon to leave, wished to see Dadabhai before leaving.[21] On his second visit, in 1893, he had a hero's reception when he arrived in Bombay, and when, a few days later, he journeyed to Lahore to preside over the Congress, he received ovations at each station from enthusiastic crowds.

It seems that Dadabhai had some idea of trying to enter Parliament as early as 1872. After the establishment of the East India Association, the subject of Indian representation in Parliament had been discussed before it more than once; and it was a natural idea that if India could not send representatives to Parliament, some Indian should try to enter Parliament as representative of some English Constituency itself. The idea had to be given up in 1872 for want of the expected support, and in 1881 Dadabhai had returned to India, as he then thought, for good. But the idea appears to have recurred towards the end of 1884, perhaps owing to a suggestion from Martin Wood[22] and in 1885 Dadabhai and Malabari began to give it a serious thought and to canvass for support. Ultimately Dadabhai left for England in March 1886, with the object of trying to enter Parliament if possible but in any case to work for certain political reforms for India, either from within or out of Parliament,[23] because it was Dadabhai's conviction that the labours of the Congress could not bear any fruit unless there was somebody in England to work for and support it. Even if he failed to enter Parliament, he could still address English audiences on India's needs and aspirations, and create English opinion in favour of Indian reform.[24]

Not all Bombay leaders concurred with Dadabhai. "From this distance it is difficult to judge", wrote Wedderburn from Bombay but for the most part we regret your determination to remain in England. There is so much work to be done here and no one so well fitted to do it as yourself. "Mehta and Telang do not like the idea of your staying in England", wrote Malabari[25], and again, "Telang has no belief in your mission to England. You are wanted here, he says." And as for Pherozeshah, "he showed himself more film even than Telang in his conviction".[26] Malabari himself felt that in the absence of Dadabhai, public affairs were not going on well at home, and splendid opportunities were going by for lack of organized action, which Dadabhai alone was capable of supplying.[27] "After all, remember," he went on to say, "You have to create, gather together, and utilize forces which must generate here, though their final destination is England."

But staying in England to try for a seat in Parliament and to do political propaganda for India required money. Dadabhai was prepared to solve his labour and the rest of his active life to these objects, but the means, he said, must be supplied by those who had them.[28] He looked to the Princes of Kathiawad in particular, to supply these means, and his friend and admirer, Malabari tried to raise funds from other sympathetic friends offering to pay upto Rs. 15,000 himself.[29] Dadabhai was anxious that the contributors should be left under no misunderstanding as to the application of their contributions. They should not think that no money was to be spent by Dadabhai upon himself. In a sense, the money would be spent on himself, but it would be for doing the public work. "Our friends do not quite understand my own personal sacrifice... I have sacrificed my means in Bombay to a considerable extent ...But it is no use talking this to others.... If Sir Dinshaw thinks that he has made any mistake in making his contribution. I have no objection to his withdrawing it."[30]

As Sir William Wedderburn was to retire from service in May 1881, a fund was started in Bombay to raise a memorial to him. Dadabhai welcomed the movement, and desired that the matter of his own fund should be held over for a time, so that it might not interfere with the Wedderburn memorial.[31]

Immediately on arrival in England, Dadabhai set about interviewing people with a view to enlisting their support for his object. (A full account of these interviews is given in the diary which he kept with him, no doubt as an aid to memory to enable him to keep his family and friends in India informed about the outcome of his efforts). Knowing fully well that "to represent an English constituency requires a good deal of preparation on English questions,"[32] he took Mr. D. Havelock Fisher's help to study the pros and cons of the various political questions of the day.[33] He knew also the value of the art of public, speaking and was anxious to cultivate it by studying some books on it. His first electioneering effort, when he stood for Holborn as a Gladstonian Liberal, was not, and also not expected to be successful. Still, he did better than could have been expected from his brief preparation, and won the respect of his opponents by the manner in which he had fought. It was necessary after this to find some other constituency. Central Finsbury was finally fixed upon by August 1888. The election came nearly four years later, and these four years were spent by Dadabhai partly in nursing the constituency, and partly in a heroic struggle against his opponents in the Liberal party itself in the Constituency. How he was selected by the Council of the local Liberal Association in the beginning, but how those opposed to him contested the selection and put up a rival candidate; how he shot into temporary fame towards the close of the year (1888) owing to an infelicitous remark of the Prime Minister, Lord Salisbury, about him: how, when the rival candidate retired and Dadabhai's friends breathed more freely, another rival candidate was put up by the group opposed to Dadabhai: how the Liberal Headquarters would do nothing to support him, but suggested arbitration, which Dadabhai indignantly refused, and declared his determination to go to the poll, whatever happened; how this second rival candidate withdrew only a few months before the election, thus leaving Dadabhai the sole Liberal candidate in the field; and how Dadabhai won the seat by the narrow margin of 3 votes; all this has been succinctly described in the introduction to the section on Parliamentary Elections.

For Dadabhai, India's honour was involved in the struggle. Having gone so far and created an interest here, he wrote to Malabari on 2nd November 1888, "... it is entirely desirable, for the honour of India, that I should spare no exertions or money to fight through. It is a trial for India herself to show whether the Indians have any backbone or not."[34]

In 1892 and 1893 Dadabhai was at the pinnacle of his fame and popularity. The news of his election to Parliament in July 1982 sent a thrill through India, and when he came to Bombay in December 1893, to preside over the Lahore session of the Indian National Congress, he was received with a rapturous welcome. His steadfastness to patriotic duty shone all the brighter against the dark background of the heavy blow that Fate had dealt him in his private life, by cutting off his only son in the prime of youth.

During his three years in Parliament, Dadabhai worked, zealously for his, constituency, and was instrumental in getting two important concessions for India: a resolution in favour of Simultaneous Examinations passed by the House of Commons on June 2, 1893, and a Royal Commission (with Lord Welby as Chairman, and with Wedderburn; Caine, and Dadabhai himself among the member) appointed in 1895, to inquire into Indian expenditure.

At the general election of 1895 Dadabhai stood again for Central Finsbury, but was defeated in common with many Liberal stalwarts. For the next five years he was busy with the Welby Commission. He submitted several Memoranda to the Commission and took the unusual course for himself by appearing as a witness before it, to be cross examined by the members of the Commission on the views put forward in his Memoranda.

During the period of his membership of Parliament and of the Welby Commission, Dadabhai was of course regularly in contact with the work of the Congress, being often consulted on the choice of the President, and also with the work of the British Committee and the Journal India. He also gave evidence before the two Indian Currency Committees those of 1893 and 1898.

After the work of the Welby Commission was over, Dadabhai remained in England for another 7 years, hoping to get into Parliament once more. When the election of 1900 came on, he was too ill to stand. For the next general election which came on in 1906, he made a great effort at North Lambeth, but was unsuccessful. In 1907 his health failed, and he returned to India to spend the last 10 years of his life in retirement, in the company of his family and friends.

The trip to India to preside over the Calcutta Congress cost Dadabhai dear. He had, as he himself said in a letter to his family, to pay the physical penalty of his visit to India. It has more or less exhausted the vitality of all my organs, and each one is taking revenge. At first there was pain in the back, which however, did not last long, but then he was unable, owing to hydrocele trouble, to take his usual walking exercise, and that, he thought, made the muscles stiff and led to pain in his feet. This ailment persisted, and to make matters worse, also suffered due to coughing in London for some time. Illness made him even give up reading newspapers. Change to the sea-side was tried; it produced some improvement; but as the ups and downs continued, it was decided that Dadabhai should return home, to spend the last years of his life among his friends and countrymen. He arrived in Bombay towards the middle of November 1907, and stayed at Versova, with his younger daughter and son in law H.M. Dadina, until his death on 30th June 1917.

At Versova, Dadabhai's health improved a little, in spite of some temporary setbacks and depression due to financial worries. He lived a retired life, never appearing in public (the only exception he made to this rule, it seems, was when he received the LLD degree from the University of Bombay on January 18, 1916) and never engaging in public controversies. People approached him with all sorts of requests like opinion on some book, advice on the redress of some service grievance, pecuniary help, his presence at some public meeting etc. To all these Dadabhai replied, excusing himself on grounds of health or of want of resources.

Apart from unselfish patriotism, the most outstanding features of Dadabhai's character were his transparent honesty, fearlessness and steadfastness in the pursuit of his duty. He was never for a moment afraid of speaking out his convictions before British audiences, or before friends like Sir George Birdwood or Sir Eiskine Perry, to whom he must have known that they were bound to be unpalatable. And when he was convinced that a certain course of conduct was right, he steadily pursued it, no matter what the immediate consequences might be, because he had a magnificent faith that ultimately right was bound to win. It was this faith that sustained him in the four years' election struggle in Central Finsbury, and indeed throughout his life. In his letters to his friends and his family, he enforced the guiding principles of his life again and again. "To lead a right life is in our power; with that we can meet all troubles and surmount and survive them." "Prosperity has not elated me and I hope adversity will not (depress) me, as long as I can feel that I am living a life of duty, of one thing I am always certain," he writes to Martin Wood, "that truth will triumph sooner or later, and I am quite content to work under that conviction, no matter what reverses or discouragements intervened." Malabari was discouraged by the opposition of social reform proposals met with. Dadabhai wrote to him not to despair in any way about the future from the shortcomings of the present. If, such shortcomings would not recur and if progress could be always easily and smoothly made, there need be no necessity for earnest workers. Once one is satisfied that he is aiming at a good object, he must go on confidently, meet all difficulties with patience, and in time, a true and good cause is bound to win. We must remain hopeful and must go on as best we can, be our progress slow or fast. This admirably sums up Dadabhai's view of life. With this faith as his armour, Dadabhai remained unperturbed even if he was misunderstood; and did got crave for personal prominence, if only the cause was served. When Malabari reported to Dadabhai that some Hindu friends had misunderstood him, thinking that as a Parsi he was bound to be loyal to the British, Dadabhai asked him not to be concerned about the misunderstanding. This happens sometimes. But truth will never die and will come out at the end. And when Malabari suggested with reference to the address to Lord Ripon, by the Press, that his own name should go on the address immediately next to Dadabhai's, which should be either first or last, Dadabhai replied that he did not care a fig where his name appeared.

It was these qualities of character that attracted to Dadabhai the respect and admiration of men so diverse as Birdwood and Hyadman, Tilak and Gokhale.

REFERENCES

1. Patwardhan, R.P. (ed.), *Dadabhai Naoroji Correspondence, Child and Youth Section,* Vol. II, Part I (Bombay: Allied Publishers), 1977, p. xiv.
2. Naoroji, Dadabhai, *Speeches and Writings of Dadabhai Naoroji* (Madras: G.A. Natesan & Co.), n.d.
3. Masani, R.P., *Dadabhai Naoroji* (New Delhi: Publications Division, Ministry of Information and Broadcasting, Government of India), 1960, p. 23.
4. Majumdar, J.K. (ed.), *Indian Speeches & Documents on British Rule 1821-1918* (London), 1937.
5. Patwardhan, R.P. (ed.), *Dadabhai Naoroji Correspondence, Child and Youth Section*, Vol. II, Part I , *op. cit.*

6. *Ibid.*
7. *Ibid.*
8. *Journal of East India Association*, Vol. II, p. 193.
9. Parekh, C.L, *Essays, Speeches, Address and Writings (on Indian Politics) of the Hon'ble Dadabhai Naoroji* (Bombay: Asia Publishers), 1887, p. 123.
10. *Ibid.*, p. 153.
11. *Ibid.*, p. 485.
12. Patwardhan, R.P. (ed.), *Dadabhai Naoroji Correspondence, Child and Youth Section*, Vol. II, Part I, *op. cit.*
13. *Ibid.*
14. *Ibid.*
15. *Ibid., Section on Voice of India*, 17-06-1885.
16. Masani, R.P., *Dadabhai Naoroji, Malabari Section* 26-12-1883, *op. cit.*, p. 80.
17. *Ibid., Ripon Memorial Section*, 22-10-1884 and 24-10-1884, pp. 80-81.
18. *Naoroji, Dadabhai, Correspondence, Child and Youth Section*, Vol. II, Part I, *op. cit., Miscellaneous Section*, 15-04-1884.
19. *Ibid., Section on Indian National Congress* 07-11-1885.
20. *Ibid.*, 24-12-1885.
21. *Ibid., Malabari Section* 07-02-1887.
22. *Ibid.*, Wood 12-09-1884.
23. *Ibid., Section on Parliamentary Elections* 20-03-1885.
24. *Ibid.*, Wedderburn 20-08-1886 etc.
25. *Ibid., Malabari Section* 30-09-1886.
26. *Ibid.*, 07-03-1887.
27. *Ibid.*, 03-04-1886.
28. *Ibid., Parliamentary Elections Section* 25-01-1885.
29. *Ibid., Malabari Section* 30-09-1886.
30. *Ibid*, 30-09-1887.
31. *Ibid.*, 06-05-1887.
32. *Ibid.*
33. *Ibid., Parliamentary Elections Section*, 05-06-1886, 07-06-1886, 15-06-1886.
34. *Ibid., Malabari Section* 02-11-1888.

2

DADABHAI NAOROJI—FROM ECONOMIC NATIONALISM TO POLITICAL NATIONALISM

An idea was conceived in some quarters that some Indian members should enter the House of Commons to carry to the mother of Parliament the hopes and aspirations of Indian population under the British Rule. It is noteworthy that before the foundation of the Indian National Congress the idea was placed before the members of the Bombay Presidency Association. At its meeting, held on 29th September 1885, there was a discussion on the subject "India's Interest in the General Election." Explaining the value and urgency of the Indian Representation in the British Parliament, Dadabhai himself laid a great stress on its need. He remarked: ". . . the present occasion of the new elections is one of those occasions in which we can appeal to the whole nation and specially in a way most useful for our purpose. It is in the Parliament that our chief battle has to be fought. The election of its members, especially those who profess to speak on India's matters, requires our earnest attention, and we should point out to the electors, which of those candidates who make India plank in their credential, have our confidence."[1] After this meeting he was convinced that he should contest for a seat in the House of Commons, although he did not disclose his decision. However, a native paper disclosed his intention when it said that Dadabhai intended to leave India for England with a view to standing 'as a candidate for election to Parliament.' It further added that "there was none equal to him in all India."[2] He was very emphatic in pointing out that not a single genuine voice was there in Parliament to acquaint it with Indian opinion on any question.[3] He repeatedly urged the need of Indian representation in Parliament.[4] Dadabhai reached England to prepare himself for entering the British Parliament, 'a body where the most fundamental questions regarding Indian administration were decided and where there was no Indian to put forth the native views on this question.' He was accepted as a liberal candidate for the Holbourn constituency in 1886. He spared no effort in his election campaign but suffered a defeat, polling 1950 votes against his opponents 3,651. Though he failed to get into the Parliament, yet it did not dampen his spirit and courage.

After his defeat in 1886, Dadabhai and his colleagues began to work for the next election with a bigger and greater determination. It was generally held that Dadabhai's entry into the Parliament would be the success of Dadabhai's 'Patriotic efforts'. One of his sympathisers wrote in 1891 "...it is necessary we should once more urge his claims upon the Liberals, and upon British electors generally, in such a manner as to convince them that India has set her heart on his return to the Imperial parliament. We must appeal to the sense of justice of the Liberal party, and ask them to be true to their principles and professions. This is a just time for united appeal."[5]

Consequently, in 1892 the executive of the General Council of the London Liberal and Radical union, of which John Morley was the president, adopted him as a liberal candidate for Central Finsbury constituency, hitherto represented by a conservative.[6] Now Dadabhai had high hopes of success with complete unity in the Liberal camp. Mr. R.M.H. Grifth acted as the chief election agent for the work entrusted to him.[7] Soon, after being adopted as the liberal candidate for Central Finsbury, Dadabhai started his election campaign. In his election manifesto he made it clear that he was a Home Ruler, not one who would give his silent vote in the House of Commons, but one who would work actively for the cause. He also announced that he was very keen on getting certain measures for reforms in the administration of Indian passed in the House of Commons. He was determined to pursue the Indian cause, come weal or woe, never deviating an inch from the singleness of purpose nor from rectitude by which he proposed to achieve his aim.

Dadabhai Naoroji himself said, "It is India's cause and I am the instrument chosen."[8] It is interesting to note that along with Indian and British supporters, Mrs. Bradlaugh Bonner, the widow of Charlie Bradlaugh great champion of Indian cause worked hard for Naoroji's election and addressed meetings in his support. Another great personality who rendered much help to Naoroji's election was his Highness Geakwar of Baroda.[9]

As a result of his five years industrious work Naoroji came out victorious in July, 1892 in the Parliamentary elections, though by a narrow majority of three votes.[10] However, Naoroji had to face tremendous difficulties in the course of his election campaign. Efforts were made to stir up racial prejudices.[11] The unity of the Empire was stimulated, and official machinery was extensively used to defeat Dadabhai. Moreover, fractions were created in the Liberal Party and the Conservative element made every possible effort to get Dadabhai defeated.[12]

The success of Dadabhai Naoroji at the poll was a matter of great joy in Congress circles. The British Congress Committee at its meeting adopted the resolution, "That this Committee desires to record its great satisfaction at the return of Mr. Dadabhai Naoroji to Parliament by the Central Division of Finsbury and heartily congratulates the enlisted electors of that constituency on being the first to recognise paramount claims of the people of India to representation in the Imperial Parliament."[13] In India his victory was hailed with universal rejoicings and the voters of Finsbury mere praised for their impartiality in returning Dadabhai 'a native of India to represent them in Parliament in preference to a countryman of their own'.[14] No other Indian at that time could have adorned that position with universal admiration as Naoroji. Lokmanya Bal Gangadhar Tilak remarked in his paper *'Kesari'*: 'If we twenty eight crore of Indians were entitled to send only one member to the British Parliament, there is no doubt that we would have elected Dadabhai Naoroji unanimously to grace that post".[15] The Eighth session of the Indian National Congress held in 1882 at Allahabad put on record 'its high esteem and deep appreciation of the services' which Dadabhai Naoroji had rendered to India, and reiterated 'its unshaking confidence in him' and looked upon him as India's representative in the House of Commons.[16] The hope was held, while the Congress would keep up the pressure for reforms, Dadabhai should keep on trying to persuade Englishmen to agree to Indian demands.

Dadabhai Naoroji's action was now to represent a constitutional gesture of early Indian nationalism at its best. He was as much a well wisher of the British government as of the Indian people. An Indian, he was the representative of British electorates: A leader of the Indian National Congress, he was a member of the British Parliament. He was to act as a link between the political circles in India and England. His main object was 'to establish a bridge of understanding between the two countries.'[17] It was Congress constitutionalism at its best. Naoroji's first effort as the member of the British Parliament was to create among members of the Parliament interest in the Indian affairs. In 1893 Sir W. Wedderburn was elected to the House of Commons representing Baniffshire constituency. It coincided with Dadabhai's membership of Parliament from Central Finsbury. 'Membership of Parliament of these two leaders strengthened the position of Indian movement inside Parliament. They were joined by the other British Liberals like John Bright, W.S. Caine, Samuel Smith, C.E Schwamm and Herbert Paul—all the supporters of Indian cause in the British Parliament. In June, 1893 the supporters of the Congress succeeded in carrying through a resolution for holding Simultaneous Examinations of the Civil Service. Naoroji and W. Wedderburn obtained a list of 130 members who readily agreed 'to take concerted action in matters relating to India in Parliament'. Several members of Parliament expressed their willingness to take up particular questions when the debates on Indian affairs came upon in the House.[18]

Encouraged by the success and sympathetic attitude of British members of Parliament, Dadabhai shared the responsibility of Sir W. Wedderburn in organising the Indian Parliamentary Committee. On July 27, 1893 a dinner party was arranged for a few independent members of Parliament. On this occasion a discussion was held in which Indian affairs were the main points. It led to moving a resolution for the revival of a committee[19] for the purpose of "promoting combined and well directed action among those interested in Indian affairs." By another resolution the committee was set up 'to promote in advance the Indian cause in Parliament'. On July 27, 1893 Sir W. Wedderburn became its Chairman and Herbert Robert was appointed its Secretary. The other members of the committee included J.E. Ellis, J. Bright, W.A. Hunter, Illingworth, Wilfred Lawsan, Herbert Paul, Joseph Pease, R.T. Raeid, G.E. Schwann, Alferd Webb, Dadabhai Naoroji, W.S. Caine, H.T. Wilson, J.G. Swift Mc Neil, S.B. Mac Loren and Walter.[20]

Due to the combined and continued efforts of Dadabhai and W. Wedderburn, the strength of the Committee rose to 152 members in 1894. But in 1895 the number reduced to 85 as sixty members suffered defeat at the general election, and several members died or resigned. However, in 1895 some new members joined on instance of Dadabhai and the total strength rose to 125. Explaining the object of the Committee Dadabhai Naoroji said that it was simply 'to take charge-of Indian questions in Parliament and to press on the House of Commons what we conceive the right views on these matters.'[21]

The Committee proved of immence value to Dadabhai in his endeavour to get Indian questions raised and passed in the British Parliament during his Parliamentary career between 1892 and 1895. The Indian Parliamentary Committee and its members kept themselves informed and in close touch with Indian leaders. Their speeches in the House of Commons were fully reported in the press. Its members exerted pressures on the British Government for the redress of Indian grievances and advocated for Indian claims. It rendered valuable and substantial services in advancing the Indian cause on the floor of the House of Commons. Telling upon the importance and success of the Indian parliamentary committee, India, an official organ of the British Congress committee remarked: "With such a committee and the larger committee of Parliament which it is, but the nucleus Indian affairs promise to receive in Parliament such an account and such a kind of attention as they never received before."[22]

The parliament was an important agency by which the Congress leaders sought to carry on their constitutional agitation in England. This kind of activity was considered to be most important by them. "India was, in a pre-eminent sense, a House of Commons concern".

The relationship between Parliament and India was considered to be "a doctor and his patient", and the doctor could make "a correct diagnosis and find out the remedy so far as to ensure a satisfactory cure.[23] In a message, issued to Indian people, Naoroji said that the battles 'of grievances and necessary reforms had to be fought in the Commons."[24] He was very emphatic in pointing out that not a single genuine voice was there in Parliament to acquaint it with Indian opinion on any question.[25] He repeatedly urged the need of Indian representation in Parliament.[26]

In July 1892, among the Indian sympathisers in the House of Commons, Dadabhai took his seat. Dadabhai Naoroji, whose action represented the constitutional gesture, was as much a well-wisher of British government of the Indian people. An Indian, he was the representative as of a British electorate. A leader of the Indian Congress, he was a member of the British Parliament. A link between the political circles in India and England, his main object was to establish a bridge of understanding between the two countries. As a Congress President in 1893, he made his principles known:

"Were we enemies of a British Rule, our best course would be not to cry out, but remain silent, and let the mischief take its course till it ends in disaster as it must. But we do not want that disaster and we, therefore, cry out, both for our own sake and for the sake of our rulers, this evil of poverty must be boldly faced and remedied."[27]

Dadabhai at that stage also felt like some others that a Congress session should be held in England. "I can assure the Congress," he said "that as I hope and wish, if you will pay an early visit to the United Kingdom and hold a session there, you will obtain a kind and warm reception from its peoples. And you will by such direct and personal appeal to the British Nation, accomplish a vast amount of good".[28]

In a parliamentary and constitutional manner Dadabhai acted as a devoted and sincere leader of India in the British Parliament. He lost no opportunity in raising his voice on Indian questions and stood out as the champion of the Indian cause in the House of Commons. He fought the battle successfully for the economic and political advancement of India, and struggled hard with large and deep issues concerning Indian administration and problems. He delivered his first speech in the House of Commons on August 9, 1892 in the course of the debate on the Address of Her Majesty the Queen. At the very outset, he pointed out, though it was unwise on his part to have stood to speak before the House of Commons so immediately after big admission to Parliament, yet he justified it on the ground that he was forced to do it under a certain necessity. Calling his election to Parliament as a unique event in the history of Indian and the British Empire as well, he spoke "in analysis of that great and wonderful phenomenon.", He called that "from the commencement of the 19th century when they seriously took the matter of Indian policy into their hands", the Britishers decided that India was to be governed on the principle of liberalism, freedom and justice. And they were doing their best to rule India in accordance with that spirit.[29] He then thanked the British people for having made possible for an Indian like him to become a member of British Parliament and sought the help and cooperation of the members of the House in the following words:

"So long as India is satisfied with the justice and honour of Britain, so long will her Indian Empire last, and I have not the least doubt that, though our progress may be slow and we may at times meet with disappointments, if we preserve, whatever justice we ask in reason we shall get.... There will be certain Indian questions, principally of administration, which I shall have to lay before the House, I am quite sure that when they are brought forward they will be fairly, considered, and if reasonable, amended to our satisfaction."[30] Dadabhai Naoroji's speech was listened to with close attention, and at the end, members of Parliament flocked round him to offer their felicitations.[31]

The question of Indianisation of Civil Services was a significant issue closely related to the Congress demand for their political advancement of the country. From the very beginning the Congress demanded association of a large number of Indians in the public services, and the holding of Civil Services Examinations in India and in England simultaneously. The demand was repeated year after year to influence British opinion on the issue. Consequently, Dadabhai Naoroji decided to raise the significant issue in the

Parliament. This was a subject that Naoroji specially made his own, and one of the points upon which he felt most keenly.[32] With the help of his British liberal friend Sir William Wedderburn, Dadabhai projected an agitation for the Civil Service Question. He tabled a bill providing for the Simultaneous Examinations in India and Great Britain of candidates for the Indian Civil Service on March 1, 1893, but the Bill was, however, dropped due to lack of support in the House of Commons. Again, he raised this question on 13th April in the House of Commons, but without any result whatsoever.[33] Undaunted by these failures Dadabhai continued his efforts and succeeded in obtaining the support of his English colleagues in Parliament like C.E. Schwann, W.S. Caine, Birkmire. On his persuasion, Sir Herbert Paul, a member of the Liberal Party moved a resolution on holding simultaneous examinations in India and in England on June 2, 1893.[34] With this the question of the Indianisation of the Civil Service assumed sudden importance. Naoroji, while supporting the resolution of Paul, charged the Government of "having taken every means in their power of breaking to the heart" the words of promises and pledges from time to time. He said: "The application to natives of the competitive system as conducted in England, and the recent reduction in the age...are all so many deliberate and transparent subterfuges.[35]

The labour of Naoroji and his other British friends bore fruit, and the resolution was carried in somewhat thin House, securing a narrow majority. Eighty-four members voted for the resolution and 76 against it. The House then resolved: "That all the competitive examinations heretofore, held in England alone for appointment to the Civil Services of India, shall, henceforth be held simultaneously both in India and England, such examinations, being identical in their nature, and all who compete being finally classified in one list according to merit."[36] This was in spite of the efforts made by the government to prevent its supporters from voting for the motion.[37] In opposing the resolution, George Russell, the Under Secretary of State for India, employed all such arguments as had been advanced by its opponents earlier. He dwelt at length on the depressing effects of this resolution and declared it not to be a wise step in regard to a sound administration.[38] Lord Cross expressed the opinion that

Government should have at once taken step to rescind the resolution.[39] Salisbury remarked that he could not imagine any project more fatal to the Indian Empire than that of identical examination.[40]

It may well be imagined how difficult it was to obtain the concession, which was entirely based on justice. The moment chosen by Dadabhai for raising this issue, was most opportune as Home Rule for Ireland movement was at its climax. On adoption of the resolution by the House of Commons, it was hailed as a 'single victory' of the principle for which educated India had been endeavouring. At the same time the resolution was considered 'a crowning personal triumph' for Naoroji a happy issue from a battle which he fought with unflagging zeal.[41] But since the attitude of the British was of indifference and of hesitation, the resolution did not receive the approval of the Joint Secretary of State for India, H.H. Fowler. He sent the resolution to the Government of India for its careful consideration and report. The Government of India characterised the proposal as 'ill admired and dangerous.'[42] The general enthusiasm, thus, received a set back to learn that the resolution was nothing more than snapping a vote. Dadabhai Naoroji deplored the British policy and attitude of the British officers. He accused the government: "you add insult to the injury. After stultifying our growth, our mental and moral capacity we are told that we are not capable."[43] "Even in India, the official attitude was viewed with great anxiety and dissatisfaction. However, Dadabhai did not give up the question. He issued letters to Indian leaders advising them to hold large public meetings at which resolutions should be passed expressing gratitude at the decision of the House and praying to the Government of India and the Secretary of State to give early practical effect thereto."[44] Consequently, floods of public meetings came into the country. They were held almost in all the important cities of India in which the Government of India and the Secretary of State for India were urged to implement the resolution of the House of Commons.[45]

In every regime, even in the most oppressive one, there might be some people or groups who benefit from the existing system and flourish. The same was the case during the British period. There were many rajas, maharajas, nawabs, zamindars, talukdars, mahajans and sahukars who lived in full satisfaction and without the least anxiety. But was the nation as a whole happy and prosperous? Clearly no, because though a few classes enjoyed, masses suffered. It was sectarian economic appeasement. The country lacked the environment of economic nationalism, in which the whole nation could enjoy the fruits of progress and production. In the words of B.G. Tilak, "A country cannot be said to have economically speaking improved so long as the conditions of the toiling majority in that country have not improved."[46] The Indian nationalist leaders, specially during the last quarter of nineteenth century, made their focus of attention the condition of masses, and not that of classes. Expressing the same spirit, the famous educationist reformer Syed Ahmad Khan (1817-1898) once observed, "To me the national honour refers to a state in which the whole nation is satisfied, majority is equipped with skill and knowledge, and all existing sciences and technology, industries and inventions are found in the nation."[47]

Dadabhai Naoroji was among the leading nationalist leaders who aroused the feeling of economic nationalism and propagated for it. The most instrumental in this regard had been the theory of drain. In the following pages we would like to study the theory and its role in awakening the desire and movement to achieve economic nationalism. We shall also examine the stages through which Dadabhai passed from economic nationalism to political nationalism or the self-rule which was his final call. The paper will conclude with a remark that economic nationalism and political nationalism are complementary and supplementary to each other and none will be realised in true sense of the word without the achievement of other.

The theory of drain is not the invention of Dadabhai Naoroji. About hundred years ago in 1776 Adam Smith in his work *The Wealth of Nations* described the British rulers of India as 'plunderers of India'.[48] In 1857 Karl Marx used almost the same words as used by Dadabhai Naoroji to describe the drain.[49] Dadabhai himself refers to about a dozen of Britishers among his predecessors and contemporaries who vouched for drain.[50]

In the 19th century the drain of wealth from India to England took the form of unrequited surplus of export over import. It was in 1867 that for the first time Dadabhai Naoroji in his paper 'England's Debt to India' put forward the idea that Britain was extracting wealth from India as a price of her rule in India, that 'out of the revenues raised in India, nearly one-fourth went clean out of the country and was added to the resources of England', and that India was consequently 'being bled'.[51]

Dadabhai Naoroji dedicated his life to propagation of the drain theory and to launching a roaring campaign against the drain which was considered by him to be the fundamental evil of British rule in India.[52]

Since the statistical methods and standardised techniques of national income calculation were not very developed at that time, there was difference of opinion on the volume and extent of wealth transferred to England from India. But the fact that there was continuous drain of wealth from India to England was undisputed issue among the nationalist leaders at that time. Only the agents and some officials of British government did not accept the drain in toto and criticised it. But the truth could not be suppressed much longer.

Dadabhai Naoroji was an admirer of British character and system. He acknowledged wholeheartedly the British contribution to India regarding education, centralised administration, discipline, political unification of the country, railways, telegraphs, hospitals, security, etc.[53] But he could not reconcile himself to the outcome of British rule in India. To him the British rule in India was un-British in character. Hence the title of his book *Poverty and Un-British Ride in India.* He expressed his earnest desire to see in India the true British government. While addressing a meeting in England in aid of the Indian Famine Relief Fund on July 1, 1900, he said, "If it were British rule and not un-British rule which governed us, England would be benefited ten times more than it is."[54]

Dadabhai Naoroji adopted a persuasive approach to cure the malpractices of British rule and get the drain checked. He frequently referred to British government's pledges and assurances given to Indians.[55] He used to quote many British officials about the importance of India to British empire and that India's prosperity was a prerequisite for Britain's prosperity.[56] Perhaps it was his persuasive tactics that he reiterated in the 2nd session of the Indian National Congress that this Congress was not a 'nursery for sedition and rebellion against the British government' and that it was 'another stone in the foundation of the stability for that government'.[57] Again in the ninth session of the Congress in Lahore in 1893 he declared, "We, the Congress are only desirous of supporting government, and having this important matter of poverty grappled with the settled, we are anxious to prevent the political danger of the most serious order declared to exist by the secretary of state himself. We desire that the British connection should endure for a long time to come for the sake of our material and political elevation among the civilised nations of the world".[58] He also appealed to the morales of the British people and the government,[59] and emphasised on cooperation to strengthen the British government.[60]

But all this was in vain. There was no sign of change in British policy of exploitation and suppression. Realising this, Dadabhai Naoroji started expressing his displeasure and disapproval of British rule in India: "...Is it just and fair, is it British that all the cost of such greatness and glory, and the prosperity of United Kingdom should be entirely, to the last farthing thrown upon the wretched Indians, as if the only relation existing between the United Kingdom and India were not of mutual benefit, but of mere masters and slaves... ?"[61] He exposed the reality of protection provided by the British government. He said, "The way you secure life and property is by protecting it from open violence by anybody else, taking care that you yourselves should take away that property".[62]

In 1895, he clearly understood the aims and objectives of the government in colonisation of India and declared that British India was indeed the British India and not India's India.[63] But he was slow in putting forward the radical demands.

Dadabhai Naoroji exposed the danger of drain forcefully and pointed out its impact on the Indian economy. According to him, it was drain that caused and intensified the famines in India.[64] It was the fundamental cause of mass poverty.[65] The drain was not limited to that of wealth but there was political and intellectual drain too.[66]

The drain was a slow poisoning to the India's national economy. According to Dadabhai Naoroji the injury inflicted to India by earlier foreign invaders was limited, it was once and over. But the British rule in India was an unending chain of drain and exploitation.[67]

Dadabhai Naoroji opposed the opium trade of Britain with China from the Indian land. According to him this being an act of immorality covered the intensity of drain. Had it been stopped, the British government would be fully exposed.[68] The British government was behaving with India like a stepmother. The other British colonies, e.g. Australia, are advancing and flourishing, but India's condition was worsening day by day.[69]

Dadabhai spent his full energy to propagate the theory of drain. After years of reconciling effort and persuasion of British authorities, Dadabhai Naoroji was disappointed from lack of any reform on the part of British rulers and accordingly the sentiment of disloyalty crept into him whose full demonstration we find in his speeches of 1904 and 1905 in which he declared that the self-government is the only solution for India's misery. In his message at the Benares session of the Indian National Congress, he asserted, "Without self-government the Indians can never get rid of their present drain, and the consequent impoverishment, misery, and destruction."[70] Thus he was guided from economic nationalism to political nationalism and that the former could not be achieved without the latter. As early as in 1876, in his essay on 'Poverty of India', he laid stress on the fact that Britain was able to keep back a large part of India's exports, chiefly because of the political position it held over India.[71] In 1896, he wrote a letter to Welby in which in an unambiguous term he stressed that drain was all simply the result of the unnatural administration and management of Indian resources by an alien country.[72] He reiterated that Indians must have their full share of public employment and a voice in their own expenditure.[73]

In Calcutta session of the Indian National Congress, Dadabhai was overemphatic when he declared in his presidential address that all the political demands of the Indian people could be summed up in one word, 'self-government or *swaraj*', like that of the United Kingdom or the colonies.[74]

Dadabhai Naoroji struggled for this cause for the next 12 years before he died in 1917. But the revolutionary spirit with which he filled the Indian masses continued to work till Indian people achieved the freedom in 1947, and forced the sun of British empire to set down in the East. Now at the passage of more than half a century on acquisition of political nationalism, let us pause a moment and ponder whether we have achieved the economic nationalism too? Are we self-reliant in our economic matters and are our resources fully utilised for the nation? Are the fruits of our development and progress fully available to the masses or only enjoyed by some classes as it happened in the past? Is it not true that for more than fifty per cent of our population, still living, independence is not complete and the dream that the great Indian leaders saw about the future of India has not been realised yet. A state where masses are not fully satisfied and suffer from illiteracy, starvation and backwardness is always an easy prey to the machination for foreign powers and vulnerable to internal unrest. It is true, as Dadabhai envisaged, that the prosperity of masses or economic nationalism is not possible without self-rule or political nationalism. Thus, both economic and political nationalism are supplementary and complementary to each other, and this is what we have struggled for.

REFERENCES

1. *Essays, Speeches, Addresses and Writings of Dadabhai Naoroji*, p. 293.
2. *The Hindu* (Madras) March 13, 1886.
3. *Report of the Indian National Congress, Second Session*, 1886, p.56.
4. *Ibid.*, 1893, Session, p. 23.
5. Naoroji Papers, Vol. IL Letter No. 180: Letter of Behramji M. Malabari. Dec. 9, 1891.
6. '*Weekly News & Chronicle*', Jan. 10, 1892 Condensed report of the meeting reported in 'India', Feb. 5, 1892, p. 28.
7. *Naoroji Papers: Letter* dated 21st December 1891, 30th July, 1893 & 25th Jan. 1895.
8. Dadabhai Naoroji cited in the *Tribune*, December 28, 1893 Vol. XVI, pp. 2-5, Columns. 1-3.
9. Vide *India*, July 15, 1892. Vol. 1, No. l, p. 192 Micro Film, Real No. 1. Feb. 1889 Sept. 1893.

10. Naoroji secured 2956 and his opponent Captain F.T. Penton (a conservative candidate) polled 2953 votes.
11. Then the Prime Minister of England Lord Salisbury called him 'a blackman' in one of his speeches in favour of conservative Candidate: *India*, August 26, 1892, p. 14, Micro Film Reel No. 1, Feb. 1889 Sept, 1893.
12. Naoroji Papers: Vol. IV Letters No. 60 and 74, addressed to William Digby by Dadabhai Naoroji dated Jan. 24 & June 12, 1892 respectively.
13. Minutes of the British Congress Committee Meeting July 1892, Resolution 3 (the resolution was moved by W.S. Caine and seconded by A. O. Hume).
14. Confidential Report on the Native News Papers of Punjab *The Paisa Akhar* (Lahore) July 11, 1892, *The Victoria Paper* (Sialkot) July 13, 1892. *The Akbar-i-Am* (Lahore) July 12, 1892.
15. *Kesarai*, July 12, 1892.
16. Report of the Indian National Congress 1892, Resolution XVI.
17. Das, M.N., *Indian National Congress Versus the British*, Vol. 1, (1885-1918), Delhi, 1978, p. 147.
18. Contribution of British Liberals to the Indian National Movement (1880 1910), A.D. Litt. Thesis of Dr. Harish P. Kaushik submitted to and approved by the University of Meerut, p. 115 (1981).
19. In 1883, John Bright approved the organisation of an informal Indian committee having for its object to secure combined Parliamentary action in matters affecting Indian public interests and 50 members of Parliament were obtained willing to cooperate on the ground of a just and sympathetic policy towards India. Vide *India*, Aug., 1, 1893, Vol. IV, No. 8, V. 241. Micro Film Real No. 1 (Feb. 1889 Sept, 1893).
20. Minutes of the British Congress Committee, Aug. 20, 1895 Vol. Jan. 2, 1894 Dec. 1, 1896.
21. *The Tribune*, August 30, 1893, Vol. XVI, No. 69, p. 3.
22. Vide *India*, May 1894, p. 135, Micro Film Reel No. 2, Oct. 1893, Dec. 1895.
23. Wedderburn's Speech at National Liberal Club London on Feb. 18, 1895 vide *Speeches and Writings of Sir W. Wedderburn*, First edition (Madras), p. 162.
24. Vide *India*, August 1895, p. 241.
25. *Report of the Indian National Congress, Second Session* 1886, p. 56.
26. *Ibid.*, Indian National Congress Session, p. 23.
27. Dadabhai Naoroji's Speech at Lahore Session of the Indian National Congress, 1893.
28. Presidential Address of Dadabhai Naoroji. Lahore Session. 1893.
29. Parliamentary Debates (House of Commons), Vol. I-VII, Fourth Series, August 9,1892, Page 260; Also *India*, August 26, 1892, p. 218 'The Imperial Parliament.'
30. *The Parliamentary Debates* (Commons) Vol. VII, Fourth Series, August 1892, p. 261.
31. *The Tribune*, Sept. 7, 1892, No. 72, p. 1.
32. *India*: March 1, 1893, p. 65.
33. *Parliamentary Debates* (House of Commons) Fourth Series Vol. XI, 13th

April 1893, p. 202 also vide *Asiatic Quarterly Review April* 1893, Vol. V, p. 507.
34. *Parliamentary Debates* (House of Commons), June 2, 1893, pp. 132-33.
35. *Parliamentary Debates* (House of Commons), Vol. XIII, Courts Series, June 1893, p. 113.
36. *Parliamentary Papers* (House of Commons), 64, 869c. 7075.
37. Kimberlay to Lansdown, June 9,1893, *Lansdown Papers.*
38. *Parliamentary Debates* (House of Commons), Vol. XIII, Fourth Series, June 2, 103, p. 125.
39. *Ibid.*, June 2, 1893, p. 125.
40. *Ibid.*, June 6, 1893. p. 129.
41. *India*, Jan. 1, 1894. p. 8.
42. *Fowler Papers: Elgin to Fowler,*. July 17, 1894.
43. *Parliamentary Debates* (House of Commons) Vol. XVII, Fourth Series, Sept. 20, 1893, p.1762.
44. *Naoroji Papers* N-I (647)—A Circular Letter of Naoroji, July 20, 1895.
45. Records of the Government of India (Home Department), 1893, Public Register No. 6724, D. No., L. No., 1382, Public B, Oct. 93, W 215 218, No. 7073, Also, Public A. D. L. No. 1405, Progs, Octr. No. 219/221.
46. Quoted in D.K. Tahmankar, *Lokmanya Tilak* (London: George Allen and Unwin Ltd.), 1956, p. 315.
47. Khan, Sir Syed Ahmad, *The Complete Collection of Lectures and Speeches* (ed.) by M.I. Gujarati (Lahore: Nawal Kishore), n.d. p. 153.
48. Smith, Adam, *The Wealth of Nations* (New York: Cannon Edition), undated, Book V, chapter III, p. 710.
49. Marx, K., and Engel, F., *On Colonialism* (Moscow), undated, p. 143.
50. Naoroji, Dadabhai, *Poverty and Un British Rule in India*, London, 1901; Reprinted by Commonwealth Publishers, New Delhi, 1988, Introduction, pp. iii, ix.
51. Naoroji, Dadabhai, *Essays, Speeches and Writings* (edited by C.L. Parekh), Bombay, 1887, pp. 29-31.
52. Masani, R.P., *Dadabhai Naoroji: The Grand Old Man of India* (London: George Allen and Unwin Ltd.), 1939, pp. 413 14.
53. Zaidi Moin A., Dadabhai Naoroji, *Speeches and Writings* (New Delhi: Indian Institute of Applied Political Research), 1985, Vol. 1, pp. 34, 22, 67, 105, 150, 224-25.
54. *Ibid.*, p. 222.
55. *Ibid.*, pp. 73-74.
56. *Ibid.*, pp. 235-36.
57. *Ibid.*, p. 3.
58. *Ibid.*, 44.
59. *Ibid.*
60. *Ibid.*
61. *Ibld.* p. 46.
62. *Ibid.*, p. 226.

63. Naoroji, Dadabhai, *Speeches and Writings* (Madras: G.A. Natesan & Co.), 2nd Edition, undated, p. 153.
64. Zaidi, M.A., *Dadabhai Naoroji, op. cit.*, Vol. 1, p. 237.
65. Masani, R.P., *Dadabhai Naoroji: The Grand Old Man of India, op. cit.*, p. 523.
66. *Naoroji, Dadabhai, Speeches and Writings, op. cit.* Appendix, p. 3.
67. Zaidi, Moin A., *Dadabhai Naoroji, op. cit.*, p. 236.
68. *Ibid.*, p. 192.
69. *Ibid.*, pp. 220, 240.
70. *Ibid.*, 440.
71. Naoroji, Dadabhai, *Poverty and Un-British Rule in India, op. cit.*, p. 33.
72. Naoroji, Dadabhai, *Speeches and Writings, op. cit.*, pp. 316-17, 319, 329, 361, 378, etc.
73. *Ibid.*, Appendix, p. 3.
74. Zaidi, Moin A., *Dadabhai Naoroji, op. cit.*, pp. 76, 100; Masani, R.P., Independence is incomplete without the welfare and prosperity Dadabhai Naoroji, *op cit.*, p. 500; Indian National Congress, Congress Presidential Addresses, Madras, no date, p. 863.

3

DRAIN THEORY

Naoroji never wrote a systematic treatise in which all the conclusions he had reached by diligent research might have been integrated. As such we have only papers of different lengths, and written for various occasions on the broad theme of the Indian economy. It would be impossible to say with any degree of finality whether he had an abstract framework that would have fitted all the scattered ideas he had generated during his long career. Any attempt to construct a consistent theoretical model with material mined out of Naoroji's vast writings may tell more about the model builder's viewpoint than about Naoroji's thought.[1] However, this need not prevent an examination of how Naoroji"s various contributions came to be composed and for what purpose. There will no doubt be an element of selection, but this will not be solely to bring about consistency in the proposed theoretical construct. Rather the main aim is to assemble the ideas in a more understandable way without violating the ideological position Naoroji held.

Dadabhai Naoroji's major achievements can be grouped under three heads: Poverty studies; Colonial economic mechanism; and Policy for development. These are not exclusive categories, but remain parts of an unmistakable pattern that defines Naoroji's economic writings inasmuch as they follow his long political Odyssey. It is remarkable that Naoroji with hardly any guide posts to steer him through the complex processes of economic change in India was able to focus unerringly on the heart of the Indian problem. No less striking is his uncanny ability to see the complicated linkages between the actual economic conditions, the systemic reasons for the existing state of affairs, and the socio-political realities. His analysis builds on his thorough knowledge of Classical Political Economy, acute powers of observation and amazing patience in collecting and analysing masses of statistical data.

Poverty and Income Studies

It is only appropriate to begin with the primary issue of the Indian economy that troubled Naoroji's mind throughout his active life. Naoroji's major achievement was to draw the attention of everyone, including a complacent British administration to the grim reality of Indian poverty. In its annual World Development Report for 1990, the World Bank economists start with the question: What do we know about the poor? Starting with a qualitative description, the Bank economists proceed towards questions of measurement of poverty.[2] This is exactly how Naoroji's analysis moves, though it is not presented in a slick and smooth way as the Bank's document. Naoroji's reading of the more conscientious and observant British civil servants sensitized him to the extent of privation being suffered by the poor in the villages. He cites numerous such statements, for example, Bourdillon of the Madras Civil Service is cited to show how "the majority of the ryots ... are always in poverty and generally in debt....A ryot of this class, of course, lives hand to mouth. He rarely sees money....His dwelling is...far ruder, smaller, and more dilapidated" than those of the far fewer but relatively better off ryots. The food he and his family consumes "is partly thin porridge made of the meal of grain boiled in water...with little condiment" from the "coarsest earthenware." This landless labourer is in an even worse condition both physically and mentally. A similar opinion given by A.F. Halsey of the United Provinces government service informed Naoroji that "the abject poverty of the average cultivator of this district is beyond the belief of anyone who has not seen it.... It may appear to many to be exaggerated, and from the nature of the case, it is of course impossible to produce figures in support of it; nevertheless, it is the result of my personal observation".[3]

Naoroji's perception of the poverty problem of India was similar, but how to demonstrate it to the attention of the British public?

Obviously if statements of Bourdillon or Halsey were produced, they could easily be rebutted flaunting statements by other less observant or partisan bureaucrats of the British Raj. Hence Naoroji set about his forays into statistical studies of Indian poverty. Naoroji was embarking on an uncharted, even hostile territory. Economists of the Classical and Neo-Classical tradition were not enthusiastic about statistical study of economic phenomena. It all began with Adam Smith himself underplaying the value of quantitative estimates of economic magnitudes. Though he contributed much to clarify key national income concepts, his suspicion of "political arithmetic" as economic statistics was called, limited its use by the dominant school of political economy in Britain.[4] This is not to deny that a lot of activity in the collection of social and economic statistics was present in 19th century Britain, thanks to institutions like, the Koyal Statistical Society and individuals like F.M. Eden, Patrick Colquhoun, Joseph Lowe, George Porter and Charles Booth to name a few outstanding figures.[5] Their labours were not given any special importance as it was accorded to the philosophical approach of Smith, Ricardo and Mill tradition. But what was quite common in the polemical pamphlet literature of the period was the extensive use of official and to some extent unofficial data. Almost every such author used data in simple ways although without any consensus about methods of analysis. If economic and social statistical enquiry was not the main player in the intellectual scene, statistical analysis of poverty was even rarer to provide Naoroji any direction. Naoroji's problems were compounded by inadequate official data, and when available, by the errors of computation he discovered. As he remarked: "I sincerely trust that future reports will not only work averages upon correct Principles, but also work out the total production .... Then only we shall know the actual condition of the mass of the people. All "I think" and "My opinions" are of no use on important subjects."[6]

Naoroji defined poverty in starkly simple terms as inability to secure a minimal standard of living. Simple as this definition is, it raises questions such as: How do we measure the standard of living? What is meant by minimal standard of living? and What is the extent of poverty? Without going into the analytical subtleties of these questions, he plunges into the question of the minimal standard of living with the words: "I now consider what is necessary for the bare wants of a human being to keep him in ordinary good health and decency." In the absence of detailed sample surveys, and elaborate consumption enquiries, Naoroji took recourse to official reports on cost of living of labourers as also of the necessary items for the survival of emigrants and prisoners under various conditions. To start with, he looks at the absolute basic minimum food required for survival "for the emigrant coolies during their voyage, living in a state of quietude" specified by Surgeon S.B. Partridge, Government Medical Inspector of emigrants. This scale, according to Naoroji "is absolutely necessary to supply the necessary ingredients of nitrogen and carbon; not the slightest luxury—no sugar or tea, or any little enjoyment of life, but simple animal subsistence of coolies."[7] From then on he moves to the cost of living in which he includes clothing and lodging for which purpose he uses the estimate provided by Kazee Shahabudin "of the lowest absolute scale of necessaries of a common agricultural labourer in the Bombay Presidency".[8]

It is now universally recognized that Naoroji's labours in preparing national income figures for the year 1867-68 initiated the work in this field, and even now no serious scholar studying economic change in modern India can afford to ignore his pioneering efforts. Partly because this was a novel enterprise and partly because he was labouring under severe handicaps, Naoroji's method was extremely simple. He took into account only the territories directly administered for which inadequate and unsatisfactory data were available. As he was convinced that the "whole produce of India is from its land", he concentrated on measuring agricultural output.

First he took the land tax collection which was around 21,000,000 pounds. As the tax represented approximately one-eighths of the gross output, he estimated it at 168,000,000 pounds. To this was added gross revenue from salt, opium and forests to make it 182,000,000 pounds. This was rounded off to 200,000,000 pounds to take into account "the produce of half a million tons of coal, of alienation of lands, or anything else there may be." This figure was further raised to 300,000,000 pounds to include "the value of manufacturing industries, excise on spirits, and a large margin for any omissions". As the population of British India was approximately 150,000,000, he calculated 40 shillings as per capita gross output of India.

Naoroji set out to improve this extremely rough and ready almost back of the envelope calculations a few years later through using more accurate and direct way of calculating the gross output and per capita income figures by using acreage, output and agricultural prices, region by region. As he ruefully remarked, he was "working with insufficient and defective materials, and without the means and time to work out details", he had to take certain short-cuts. One such was his decision to use only the "largest one or two kinds of produce of a province to represent all its produce." This meant that he took only the prices of paddy and cumboo in Madras, coarse rice in Bengal, wheat for N.W.P. and wheat and inferior grains for Punjab.

Taking the cultivated area in the districts, output per acre, and the price level, he determined the figure of average value of net income per acre for these two principal crops. This average value was multiplied by the total acreage of all crops under cultivation in the province to give the value of the produce of cultivated land. The total for the seven provinces came to 277,000,000 pounds. He does not add grazing land or straw as they are required for the cattle, but deducts 6 per cent of the total for seed next year leaving a balance of 260,000,000 pounds worth food available for "human use and consumption for a year." A sum of 15,000,000 pounds was added "as a guess only" to account for meat, fish, poultry and diary products. Another 17,000,000 pounds takes care of "salt, opium, coal and profits of commerce" but these are better than mere guesswork as Naoroji was able to get data from government sources on physical output and value of minerals produced that year. Industrial output was estimated using another convenient shortcut. The estimated annual output of Punjab in shawls, silks and other products came to 1,771,000 pounds. Making allowance for the raw produce used, he scaled the value down to 2,000,000 pounds. For the rest of the country where the manufactures "are not quite as valuable", he arrived at a figure of 15,000,000 pounds, using the figures arrived at Central Provinces as the standard. To this grand total of 307,000,000 pounds, he added a further 33,000,000 pounds for "any contingencies" to make up the total, a round figure of 340,000,000 pounds. With an estimated population of 170,000,000 he got the per capita income figure of 40 shillings or Ks. twenty. Naoroji claimed that this figure could not be an underestimation, if anything, it could have over estimated the income figures. This is so because he had not made for any allowance for bad seasons. If all that were allowed, he concluded "cannot help thinking that the result will be nearer 30s than 40s a head."[9]

Naoroji's efforts at constructing national income figures were essentially meant to highlight the poverty problem of India. During the last few decades, economists and statisticians like Minhas, Dandekar, Rath and Vaidyanathan have studied the question of the "Poverty Line" with great sophistication and ingenuity. In substance what they have done is not conceptually different from Naoroji's early labours. They have:

1. tried to determine the minimum nutrition required for subsistence;
2. estimated the cost of this minimum quantity; and
3. on the basis of the per capita consumption expenditure drawn the poverty line.

On the actual drawing of the poverty line, Naoroji did not have the benefit of National Sample Survey's percentile breakdown of consumption expenditure data or other similar surveys. Consequently, Naoroji never attempted to find out how many were actually below the minimum required sustenance. He was, however, acutely aware of this issue, as he said: "It must. . . . be borne in mind that every poor labourer does not get the full share of the average production. The high and middle classes get much larger share, the poorer classes much less, while the lowest cost of living, is generally above the average share."[10] Naoroji's thesis was that on the average nobody except the very few fortunate were above the 'poverty' line as the total national output was much less than it should be. The average value of production per head was far below the minimum standard of living, and even below what was actually spent on the prisoners in various jails.

The obvious conclusion to be arrived in Naoroji's scathing words: "Even for such food and clothing as a criminal obtains, there is hardly enough of production even in a good season, leaving alone all little luxuries, all social and religious wants, all expenses of occasions of joy and sorrow, and any provision for bad season ... Such appears to be the condition of the masses of India. They do not get enough to provide the bare necessaries of life."[11]

The Colonial Economic Mechanism

Having delineated the poverty problem of India, Naoroji's task was to identify the causes that resulted in the all round deprivation that the general run of Indian population was suffering from. While his writings attempt to give various explanations and interpretations of the miserable economic conditions prevailing, Naoroji assigns a predominant position to the nature of the colonial rule in India. At the same time, in order to condemn the British rule he does not imagine the existence of a golden age in the past destroyed by the colonisers. In fact, he has consistently upheld the value of British rule for India at least in his early years, and appreciated the educational system, the new political institutions, freedom of speech and other reforms that were introduced since the advent of the British administration. He certainly approved the bright side of British rule, and the many blessings of law and order which it has conferred on India."[12] Similarly he had nothing against British individual's official and non-official alike. It was the "system which is evil and bad" and that "makes very good men work on wrong lines."[13]. In other words, the sufferings and poverty encountered in India were the result of systemic causes. The major contribution of Naoroji to the study of Indian economy is to elaborate the structure, operation and results of this colonial economic system.

Naoroji held that "Drain" was at the very heart of the colonial economic mechanism. This is not something that a crusty nationalist politician pulled out of his hat but the result of the workings of an economic scientist's mind. A sound economist begins by adopting, on the basis of a preliminary examination of the facts, a broad prescience of the economic process. He starts by orienting himself towards some key factors which his prescience or vision informs him to be of causal significance for the structure and operation of the economic system. The next step is a thorough examination of the economic facts which should enable him to arrange the factors on a scale of importance. This will naturally depend on his initial assessment, his ideological beliefs, and political sympathies but all tempered by the uniformities and regularities that he observes in the economic details. Taking the factors that have been ranked at the top by his study of facts, the economist tries to give an explanation of particular phenomena by constructing economic models. In Naoroji's case, "Economic Drain" is the model by which he analyses the nature of the 19th century economy, and gives an explanation of the poverty observed amidst a predominant section of Indian population.

The Drain Theory suggested that the colonial economic system enabled a vast transfer of income and wealth from India to Britain for which nothing was received in return. On the contrary this continuous outflow of resources resulted in economically ruining the country. The burden of drain was not seen solely in the amount of funds transferred without receiving any benefit. Naoroji saw in this an element of opportunities foregone as he wrote: "The loss to India must be measured by how much more India would have benefited, had this enormous drain of the ten years and all drain of previous years been at India's own disposal and fructified, in the Indians' pockets."[14] The drain was not merely to be equated with one time plunder, but something far more insidious and deadly. Naoroji making a reference to the 11th century Afghan invader of India, Mahmood of Ghazni said that at least after the 18th invasion the plunder stopped, but this British colonial drain seems to have no end.[15] The process of drain, he argued, is so invisible that one doesn't realise its existence.[16] Though there is a conceptual clarity in the "Drain Model", Naoroji's conclusions are not systematically presented. Taking an encyclopaedic view, he had tried to observe all aspects of the colonial rule by adopting a comprehensive and broad definition of the phenomenon of Drain.

The colonial economic mechanism rests on India being administered by foreigners and their administration was quite insensitive to the needs and aspirations of the country. Unlike other colonies like Australia and Canada which had to be developed first before extracting any surplus for the benefit of the colonising country, India's riches were already there for primitive extraction even before the establishment of the British rule. The early trader/soldier combination did drain large amounts of treasure through ignorable methods of "shaking the Pagoda Tree". That crude form and methods of draining had changed in the nineteenth century but not the fact of drain itself. The gargantuan administrative apparatus created by the colonial rulers itself was the primary source of drain by being enormously expensive. After analyzing the salary structure and employment pattern, Naoroji came to the conclusion that by merely employing Indians (let alone cutting down the number) instead of the highly paid Europeans, the government would have reduced the administrative expenditure by a quarter.

The other major source of drain which Naoroji never ceased to highlight was the immense scale of Indian public expenditure year after year. Bulk of the expenditure was incurred on the army, police, interest on debt which besides being of an unproductive nature were only serving the interests of the colonial masters and not for the welfare of the people of India. Even those areas of public expenditure which on the face of it appeared to be for stimulating economic progress such as railways, public works, telegraphs etc. were only serving the strategic requirements for maintaining the empire intact.

Both the above items were financed out of what he often called "Crushing Burden of Taxation". The system of taxation was highly regressive, and bulk of the revenue came from the rural poor.[17] This extraction of vast sums from the Indian people through taxation and spending unproductively on a flabby expensive bureaucracy, and on waging wars defending the empire for the benefit of the British Crown was one kind of drain. It may not be maladministration in the sense that it was inefficient. On the contrary, the administration's ruthless efficiency created a situation where Indian enterprise was kept down and misery increased as a result of not spending where it was needed and spending where it should not have been.

The kind of drain from the predominantly rural poor characterised by Ganguli as "Internal Drain" was bad enough, but when it was combined with an external drain, the results have been nothing short of disaster.[18] Naoroji devoted a large part of his intellectual energies to lay bare the intricate connection between the internal drain and the external drain. While in the first place, one part of the colonial economic system extracted resources from the Indian economy, subsequently another part of the system transmitted those resources to Britain. The result was a substantial one way flow of wealth from India to Britain with the former receiving nothing in return. This can be observed in the excess of exports over imports of India, and Naoroji calculates that in this manner. Britain had over the 38 year period 1835-1872 "kept back as its benefit" a sum well above 500,000,000 pounds sterling "chiefly arising from the political position it holds over India."[19]

The composition of the Drain, he said, consisted of the following:

(i) Remittances by European officials (a) of their savings, and (b) for their expenditure in England from their salaries and pensions.
(ii) Indian government's expenditure in Britain including purchase of government stores and salaries and pensions paid in there.
(iii) Remittances by non-official Europeans like planters and businessmen.
(iv) Interest payments in England for English capital (which was nothing else than what was remitted earlier) invested in India like Railways.

Having drawn a clear picture of a backward economy Dadabhai now proceeded to expose the misrule of the British rulers in impoverishing India, through his drain theory. The theory has two aspects—one is the external drain and the other is the internal drain. The external drain is related to that part of the remittance to Britain which did not bring any return to India. It was usually in the form of unrequisite exports on non-commercial exports. The internal drain on the other hand was in the form of heavy taxation without any compensatory effects of beneficial public expenditure. Dadabhai took the export import figures of India for the years 1835 to 1872 and found that the value of exports exceeded that of imports by £ 500,000,000. Had interest been calculated the drain would have amounted to a higher figure. He then concluded that most of the ills of India were due to this heavy tribute which she had to pay England and she was "verging to the lowest ebb of pauperism". John Crawfurd (1837) in his paper "A Sketch of the Commercial Resources and Monetary and Mercantile System of British India" gave in detail the charges that India was to meet from her revenue in the year 1835. Those were as follows*:

	£
Dividends of proprietor of East India Stock chargeable upon the revenues of India, Act 3 and 3, Will, IV, Cap, 85	636,826
Interest on the Home bond debt, ditto	92,858
Invoice value of stores, consigned to India	234,341
Transport of troops and stores, deducting freight charged in invoice	40,944
Furlough and retired pay to military and marine officers, including off reckonings	521,316
Payment on account of King's troops serving in India	120,000
Retiring pay of King's troops	60,000
Charged general, deducting the charges of establishments put upon outward invoices	404,890
Absentee allowances to civil servants of the Indian establishment	38,633
Miscellaneous	5,133
Total	~~2,154,941~~

All these are monetary items that show up in the Balance of Payments tables, but Naoroji added what he called "Moral Drain" by which he meant drain of experienced human resources. After

* East India Annual Revenue Accounts: Printed on 24 June 1836.

all Europeans sit in all positions of responsibility and power in India, and when they leave they take away all the "experience and wisdom" which they had acquired during their stay in India. Thus, "India is left without, and cannot have those elders in wisdom and experience who in every country are the natural guides of the rising generations in their national and social conduct."[20]

Dadabhai thought of this economic drain as an external drain, external in the sense that there was a 'drain', that is the unilateral transfer of social surplus, or potential investible capital to Britain by the colonial state, its official and foreign merchants and other capitalists through unrequited exports, 'non-commercial exports', for which India got back no equivalent economic, commercial or material returns for it in any form. The result was that through the process of external trade a considerable part of the surplus thus generated in India served as capital and entered into extended re-production, not in a low level colonial economy like India. But in Britain and its white colony, reflecting the fact that international trade then acted as an instrument of exploitation and under development. But, according to him, there was also an internal drain in the form of transfer of purchasing power through the medium of taxation largely on the poor of India, interest payments, profits and other forms of internal surplus which were utilized very little on the development of agriculture and industry or on social infrastructure in India.

He attached a number of objective factors such as the following for the well functioning of this kind of built mechanism of transfer of resources (external-cum-intemal drain):

(i) India is a vast country ruled by a few Europeans, towards the objective of opening up the country to free trade—an euphemism for exploitation and towards the objective of creating highly paid jobs for European personnel. The related income enjoyed by the latter is a 'moral drain', a cost to British India.

(ii) India develops as a market for British manufactured goods and a supplier to Britain of its food and raw materials. This is because India's policies are determined in Britain and in the interests of the British economy and the British capitalist class.

(iii) The India government is forced, under British rule, to meet an ever increasing long list of official overseas expenses. This can be described as Home Charges. (See Table on page 58).

(iv) India is colony with a difference, public expenditure out of the proceeds of taxation and loans failed to generate as much of domestic employment and income.

(v) India, being increasingly reduced to the status of a mere agrarian appendage and a subordinate trading partner of Britain, becomes a classic colony governed by remote control.

The colonial economy's structure and its mode of operation may be pictured in terms of inter sectoral relationships among its key sectors.

The hazy concept of 'Moral Drain' apart, there was certainly the question of undoubted long-term transfer of resources out of India. What was the effect of a sustained outflow over a lengthy

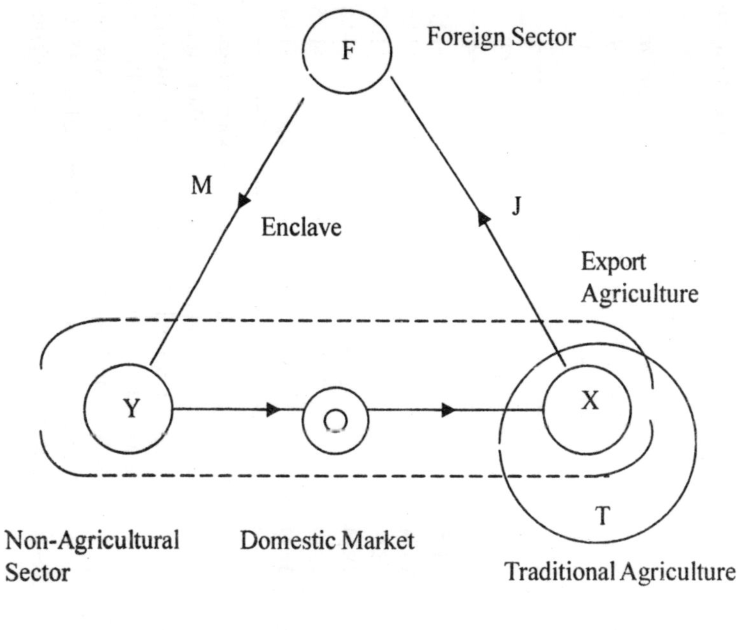

India, Annual Balance of Payments of Current Account, 1869-70 to 1894-98

(£ millions, quinquennial average)

	Balance Merchandise Trade	*Net Treasure Imports*	*Balance Visible Trade* (1+2)	*Home Charges*	*Other Invisible*	*All Invisibles* (4+5)	*Balance of Payments Current Account* (3–6)
	(1)	(2)	(3)	(4)	(5)	(6)	(7)
1869–73	+22.6	–8.4	+14.2	–8.8	–15.6	–24.4	–10.2
1874–78	+21.0	–6.4	+14.6	–9.3	–18.0	–27.3	–12.7
1879–83	+23.8	–7.1	+16.7	–10.7	–17.7	–28.4	–11.2
1884–88	+23.8	–9.2	+14.6	–12.3	–18.0	–30.3	–15.7
1889–93	+25.2	–9.7	+15.5	–13.5	–19.4	–32.9	–17.4
1894–98	+20.7	–5.6	+15.1	–13.9	–18.9	–32.8	–17.7

Note: A plus sign (+) indicates net exports of goods; a minus sign (-) indicates net imports of goods and net exports of remittances, service charges and other invisibles.

The most through direct estimate of flows of long-term foreign capital into India from 1870 to 1899 gives a total of between £ 123.2 million and £ 144.8 million, most of which was in the form of sterling loans to the Secretary of State for India in London (see Lance E. Davis and Robert A. Huttenback, Mammon and the Pursuit of Empire. The Political Economy of British Imperialism, 1860-1912, Cambridge, 1986, Table 2.1)

Source: A. K. Banerjee, *Aspects of Indo-British Economic Relations, 1858-1898*, Bombay, 1982, Tables 34 A and 40 A.

period of a substantial portion of domestic output on the state of the Indian economy? Naoroji suspected that it must have had considerable adverse effect on domestic incomes, prices, employment, capital formation and income distribution because many civil servants have given hints about such situations and Naoroji was an avid reader of government reports. There is for example a report by Richard Temple cited by Naoroji which is significant:

"In former reports it was explained how the circumstances of so much money going out of the Punjab contributed to depress the agriculturist. The native regular army was Hindustani; to them was a large share of the Punjab revenues disbursed, of which a part only they spent on the spot, and a part was remitted to their homes. Thus it was that, year after year, lakhs and lakhs were drained from the Punjab, and enriched Oudh. But within the last year, the native army being Punjabi, all such sums have been paid to them, and have been spent at home.... The effect of all this is already perceptible in an increase of agricultural capital, a freer circulation of money, and a fresh impetus to cultivation."

Naoroji extended this interregional phenomenon to the international sphere by looking at the British India Britain economic connection: "This is just the cause which, in a far more aggravated form and on a far larger scale, operates on the whole of British India in its relations with England. Millions are drained to England and till the reversing cause of the retaining and return of wealth in someway comes into operation, the evils of the drain cannot be remedied."[21]

Naoroji's intuitive understanding of what has come to be known as the transfer problem is quite fascinating. The transfer problem is an examination of how unilateral payments made from one country to another outside normal trade transactions get settled. The paying country has to generate an export surplus in order for money payment to be converted into real resources. Nineteenth century economists argued that this transfer is made either through the price-specie flow mechanism or through a transfer of demand in real terms. The latter method was of much interest to many economists because of the burning issue of absentee Irish landlords living in England and receiving payments.[22] This was a continuous process creating a long run balance of payments disequilibrium which could only be adjusted through shifts of demand.

The orthodox position in England was that the English rent receivers will increase their demand for Irish products. In other words, the money payment made would be covered by Irish exports which would after all encourage productive activity in Ireland.[23] This was criticised by contemporary economists with an Irish background like Longfield who questioned the adequacy of using shifts in demand to explain away a very real problem. Similarly, John Stuart Mill argued that unilateral transfer of funds would shift the terms of trade against the transferring country and in favour of the receiving country: "The result is that a country which makes regular payments to foreign countries, besides losing what it pays, loses also something more, by the less advantageous terms on which it is forced to exchange its productions for foreign commodities."[24] Thus to use the words of Samuelson: "in addition to the 'primary burden' on the paying country of the direct payment, there would in addition be a 'secondary burden' as her export prices deteriorated relative to her import prices."[25]

Though Mill does not refer to drain from India and mentions only the "remittances of rent to absentee landlords", he does mention "interest to foreign creditors" and "a government expenditure abroad" as two key items in the Indian drain. R.D. Collison Black has pointed out that Mountifort Longfield, his disciple Isaac Butt and John Stuart Mill were the only major economists to have effectively criticised the orthodox doctrine espoused by J.R. McCulloch.[26] It is interesting that the first two were Irish, and John Stuart Mill had almost ringside knowledge of what happened in India because of his official position at the East India Company's headquarters in England. Naoroji had read Mill's Principles of Political Economy which he often cited in his writings. Given Naoroji's activities in London, he might have even met Mill. Similarly, his stay in England must have sensitized him to the views of Irish patriots like C.S. Parnell, Leader of the Home Rule Party who were championing agrarian reforms to reduce the rent being sent to the absentee landlords in Britain. Irish nationalists were conscious of the drain of capital through increased taxation and rents sent overseas. They argued that once the drain was stopped by political reform the Irish economy would flourish.[27]

Having identified 'economic drain' as the main cause for Indian poverty, Naoroji explains how. To begin with, he disposes off the contention of some British administrators that India's economic travails stem from the operation of the economic law of overpopulation. Naoroji was not convinced and asked how could Britain not suffer despite the enormous increase in her population during the past many decades. Macaulay is brought as witness to say "that the advantages arising from the progress of civilization have far more than counter balanced the disadvantages arising from the progress of population."[28] There is nothing such as inevitable consequences of overpopulation. It was really "the pitiless perversion of economic laws by the sad bleeding to which India is subjected."[29]

It was really the lack of capital or rather the capital that was continuously being drained out of India that was responsible for Indian poverty. Naoroji's reading of classical political economy, especially Adam Smith and John Stuart Mill informed him that capital accumulation was crucial for a country's progress and not just demand. He often cited passages from Chapter V, Book 1 of Mill's *Principles on Fundamental Propositions Respecting Capital* to show that "Industry is limited by capital." Again that capital was possible only with saving. Not that he considered demand to be of no consequence but it was important only insofar as it influences allocation decisions concerning the deployment of capital.[30] Applying these ideas, Naoroji showed that when the poor in India do not have enough to meet the minimum requirements of living, where was the question of saving? Add to this the richer classes, few as they were and capable of capital accumulation consumed a "portion of the produce . . . unproductively in luxuries". On top of this, the annual unilateral transfer of resources to Britain further reduced the capital available for investment in India. The inescapable result then "is an additional evil consequence in the capability of labour deteriorating continuously for 'Industry is limited by capital' so the candle burns at both ends capital going on diminishing on the one hand, and labour thereby becoming less capable, on the other, to reproduce as much as before."[31]

If lack of capital was the biggest cause of Indian poverty, why was his opposition to railway investment in India? Naoroji made it clear that under normal circumstances, i.e. when India was not being hegemonised, he would welcome foreign direct investment of capital: "there can be no objection to any capitalist, or banker, or merchant, or manufacturer going to India on his own account and making any profits there, if we are also left free to do our best in fair competition."[32]

He was fully conscious of the benefits of public works like railways to the economy, and even highlighted how British investment in American railways had benefited the people and economy of that country.[33] Railway investment in India, on the other hand, was part of an unequal exchange underlined by an unnatural system of administration and management of expenditure. The railway capital coming into India represented but a small portion of the transfer of capital that had already taken place intended to close the gap between the transfer of tribute and exportable surplus of goods. By increasing India's indebtedness further, it was only causing further drain in the form of interest payments. The provision of guaranteed rate of return to the railway companies introduced an element of inefficiency by the Indian government bearing the risk of investment. Like the rest of the administrative machinery, railways too were manned by Europeans except at the lowest level, and naturally they repatriated all their income to their countries. In Naoroji's colourful words: "English capitalists do not merely lend, but with their capital they themselves invade the country. The produce of the capital is mostly eaten up by their own countrymen, and, after that, they carry away the rest in the shape of profits and dividends."[34] After all this, does the country receive any benefit? No, said Naoroji because railway lines were determined more for strategic and military considerations to protect the empire rather than for commercial economic developmental reasons.

Policy for Development

Dadabhai's ultimate objective was to change conditions in India for the better. As his main conclusion was that it was the colonial economic system that was responsible for the prevailing conditions through the drain mechanism, reform of the system was the prime necessity. His elaborate critical comments implied what kind of reforms Naoroji had envisaged. The fundamental requirement was to put an end to the drain *in toto*. Once the continuous flow of resources from India ended, and benign British rule established (as against the existing un-British rule), India will be able to progress and end the prevailing poverty.

The policy framework Naoroji had in mind was inspired by John Stuart Mill's views on economic development. This would include a good government, moderate and equitable taxes, security of property, freedom from arbitrary collection of taxes, a good system of land tenure to encourage cultivators to be productive, provision of "a comprehensive plan of national education" to stimulate mental and other inventive activity, inflow of the right kind of foreign capital and enterprise.[35] The primacy of place was given to capital accumulation which can take place only when the drain was plugged. The key reforms needed to achieve this was to Indianise the administration, cut down unproductive expenditure and encourage private initiative.

Firstly, Naoroji argued for employing Indians at all levels except at the very top. This should apply not only to the government sector but also private sector enterprises like the Railway companies. While the aspirations of the Indian youth to occupy positions of power was important,[36] Naoroji said that the employment of Indians by itself "is not the end or ultimate object... (it) is the only means to the great and all important end of stopping the drain, deaths, destitution, and of promoting the relief, welfare, and prosperity" of the suffering people of India.[37] Not only will greater employment of Indians prevent drain and be more economical because of lesser remuneration,[38] Indians will have greater empathy with their fellow Indian subjects. They will be what the colonial administrator is not. The Europeans, he claimed "do not belong to the people; they cannot enter their thoughts and feelings, they cannot join or sympathise with their jobs and griefs."[39] In short, by Indianising the administration, India's stock of wealth, work and wisdom will go up immeasurably.

Secondly, a key policy to reform the state of Indian finance was to prune government expenditure ruthlessly. Every item of expenditure according to him, should be assessed separately in terms of: (a) sufficiency, (b) efficiency, (c) achievement of the stated objectives "at an economical and affordable cost", (d) necessity, and (e) "whether there is any peculiar inherent defect."[40] If such an assessment were made, the biggest cut would fall on military expenditure. Like all nineteenth century economists, he was quite aware of the essentially unproductive nature of military expenditure. But there were other reasons why military spending should be pruned. Afterall as Naoroji claimed correctly, Indian government was maintaining the military machine "solely and entirely for British purposes and British interests". Having condemned military expenditure in no uncertain terms as the "very badge of slavery of a foreign domination and tyranny", still he thought it gave India some indirect benefit. India "may pay some fair share of the cost" but only after considering item by item. For example, India and England could pay all salaries "which are to be paid to their own people". But in the case of Navy, the entire cost should be borne by England.[41]

It was not that Naoroji was against public expenditure as such because he knew well enough the productive possibilities of well administered public outlay. It was his contention that "the real evil and misery of the people" was not the outcome of too much expenditure. On the other hand, "India is capable, under natural circumstances, of providing twice, three times or more the expenditure, as the improvement of the country may need, in attaining all necessary progress."[42] The problem lies in the way "that expenditure is administered and managed". Actually the real issue was "not how to nibble at the expenditure and suggest some poor reductions here and there, to be put aside in a short time"[43], but how to guarantee good governance.

Thirdly, Taxation should be moderate and fair. Naoroji's discussion of taxation was simply that portion of income extracted from the people, as he says: "It is so much taken from the income of the country for the purposes of Government". From this follows it is conclusion that though taxation per capita is higher in England than in India, percentage-wise Britain takes only 8 per cent, whereas it is 15 per cent, in India. Taxable capacity is important and it cannot be "measured by the percentage of taxation, but by the abundance or otherwise, of the means of income to pay it from." Naoroji argued that taxable capacity can be judged only by how much is left in the hands of the taxpayer after his essential requirements are met. As he explained: "From abundance you may give a large percentage with ease; from sufficiency, the same burden may just be bearable, or some diminution may make it so; but from insufficiency, any burden is so much privation."[44] From this micro-economic conclusion, he moves to a macro economic judgement arguing that the tax burden was much higher in India because a substantial portion of the tax did not return to the tax payers as they had been drained out of the system.

Quite apart from the larger issue of taxable capacity determining tax rates, Naoroji was concerned about the moral issues of taxation.[45] This comes out clearly in his total opposition to opium revenue which was one of the major sources of income for the government. Naoroji charged that while Britain curtailed the consumption of opium within, it "tempts a vast human race to use it" solely to shift some of the burden of economic drain from India to China. Though India was benefiting by the opium trade, he had no hesitation in claiming it as "a sin on England's head, and a curse on India for her share in being the instrument."[46] Similarly, he argued against salt tax as it was a tax on an absolute necessity for the rich and poor alike.

Naoroji was firm in his support for import tariff to help the nascent Indian textile industry for which he derived much support from Mill's infant industry argument. Some of his trenchant criticism was reserved for the removal of cotton duties in the mid-1870s in the name of free trade. He was not a narrow-minded nationalist holding 'beggar my neighbour' type of Mercantilist ideas. Like the British liberal thinkers, he did believe in free international trade, improving the welfare of all the trading partners. All the same, he was willing to go against free trade if the necessary preconditions for its successful operation were not present. He told the Calcutta Congress: "I am a member, and in the Executive Committee of the Cobden Club ... and yet I say that Swadeshi is a forced necessity for India in its unnatural economic muddle. As long as the economic condition remains unnatural and impoverishing ... to talk of applying economic laws to the condition of India is adding insult to injury."[47] Free trade can take place only among equals, but as he said it was not the situation between England and India. He thought that import duties on cotton textiles was a much less objectionable tax than other taxes on necessities like salt.

Fourthly, predictably Naoroji advocated reduction of public debt. Indian public debt was a political burden and forced on India "for the very acquisition and maintenance of the British rule."

Finally, Naoroji's good grasp of the workings of both the monetary system and the foreign exchange mechanism coupled with his basic orientation on the nature of economic drain, enabled him to understand the impact of monetary events on India's economic position. It was during the early 1870s that a phenomenal increase in World silver production took place. With India under a silver standard and prices of precious metals determining exchange rates, Indian Rupee suffered continuous devaluation. This had stimulated Indian exports but during the same period had witnessed significant increase in India's foreign indebtedness. Public expenditure also shot up necessitating a substantial increase in tax collected. Whereas the colonial rulers saw the problem as an inevitable consequence of the silver gold price fluctuations, and imagined that it could be solved by altering the monetary standard, Naoroji argued otherwise. It was not solely a monetary problem but something that had ramifications in Indo-British relations and economic development of India. In fact, if the existing system were altered to arrest the slide in the value of the Rupee, it would only put more strain on the tax system. As he said, "closing the mints or introducing a gold standard does not and cannot save a single farthing to the Indian tax payers." It will actually lead to higher taxation to pay India's sterling debt charges.[48] Naoroji's policy was to leave the currency system undisturbed on the existing silver standard because any other policy would give discretionary power to the government to manipulate the system only to benefit the colonial masters.[49]

The core of Naoroji's argument consists of the following propositions:

1. British conquest, though it began with the intent of exploitation, matured into a benign rule providing in the process many social and political preconditions necessary for economic growth through the spread of capitalism.
2. Instead of completing this task through the ideal British model defined by a non-exploitative rule, a colonial economic mechanism was allowed to be developed to take economic advantage of India through political domination.
3. The repressive economic administration and a massive unilateral transfer of resources made capitalist development impossible in India because it affected the one crucial element necessary, namely, capital accumulation.
4. Not only development was not possible, immiserization of the mass of Indian population became inevitable.
5. If India had to escape from poverty, the colonial economic mechanism had to be dismantled and freedom given to Indian to take economic decisions to suit the national requirements. A start should be made by Indianisation of employment in the public and quasi public sector.

This is an original and imaginative analysis as it identified the problem, investigated its causes and provided a framework of policy measures. What Naoroji accomplished is what policy analysis is all about.[50]

It would be doing injustice to claim Naoroji to be an economic scientist. He did not produce carefully structured abstract treatises, nor was he concerned to serve the discipline as such. His intentions were to influence the course of events by informing the public of the true state of affairs. At the same time, it should be understood that his writings were not ephemeral to the point of being forgotten the next day. There is a core of scientific element in his work. But this lies in his remarkable perception of the major economic implications of colonial rule. Naoroji's ability to cut through a lot of fog and get to the heart of the Indian poverty problem, and then to apply sturdy commonsense, sociological/political insights and economic wisdom to its resolution is evident in all his writings once we are able to translate them from their shrill polemical form of argumentation. It is also possible to conjecture that the "polemics" itself was part of his style of discourse.

The use of the word "Drain" by Naoroji is significant. Modern economists use the less colourful word "leakages or unilateral transfer" for such phenomena. But Naoroji belonged to the nineteenth century political economy tradition which understood the value of rhetoric in economics to persuade people. Economists of the 18th and 19th centuries clearly realized the suggestive power of terminology. From the "Stationary State" or "Invisible Hand" of Adam Smith to the "Rate of Exploitation" or "immiserization" of Marx, social scientists consciously used words with a high emotive content. Joan Robinson was not far wrong when she said: "No school of economics has ever used a perfectly colourless terminology. Overtones ring in the mind of the reader even if the writer believes himself to be coldly scientific".[51]

Naoroji's intellectual tools may appear crude, but there is no doubt that his sense of reality was very strong. What is astonishing is his uncanny ability to look at past events and draw out valuable lessons. By connecting the heavy taxation burden, the enormous unilateral transfer of capital, and problems of monetary standard, Naoroji showed an unusual understanding of the macro-economic aspects of the Indian economy for the first time in the history of economic thinking. He knew what the central problem of the Indian economic system was, and knew precisely what questions should be asked, and what kind of analytical tools should be used for conducting an economic discourse. After all, in economics it is even more useful to ask relevant questions, and give approximately correct answers with simple tools, rather than parade sophisticated techniques to answer "accurately" spurious and irrelevant questions.

One measure of relevance and importance of the contributions of an economist is the extent to which they have stimulated further work. In this sense, Naoroji's writings are as relevant and important now as they were when we pioneered the study of poverty, national income, internal and external economic leakages, problems of balance of payments, public expenditure, burden of taxation, and economic hegemony. Whatever aspect of Indian economics he touched, he has provided rich insights which have anticipated and inspired a phenomenal amount of research both among his contemporaries and later scholars.

Frank Knight, the great economist said that the "basic principle of science truth or objectivity is essentially a moral principle, in opposition to any form of self-interest. The presuppositions of objectivity are integrity, competence and humility." To this list was added another quality, namely, courage, i.e. to be fearless in the defence of the causes in which the practitioner believes.[51] These apply in full measure to Naoroji the activist political economist. Judged by what has been said above, Naoroji stands high in the ranks of the profession of economists as practised in India during the last one hundred years.

REFERENCES

1. Kurdas, Cigdern, "The Whig Historian on Adam Smith: Paul Samuelson's Canonical Model", *History of Economics Society Bulletin*, X. 1 (Spring), 1988.
2. World Bank, *World Development Report, 1990* (Oxford: Oxford University Press), 1990, p. 24.
3. Naoroji, Dadabhai, *Poverty and Un-British Rule in India* (New Delhi: Publications Division, Government of India), 1962.
4. Dimand, Robert, W., "I have no great faith in Political Arithmetick': Adam Smith and Quantitative Political Economy" in Ingrid H. Rima (ed.), *Measurement, Quantification and Economic Analysis: Numeracy in Economics* (London: Routledge), 1995.
5. Stundeski, Paul, *The Income of Nations* (New York: New York University Press), 1958, Vol. I, Chapter 7.
6. Naoroji, Dadabhai, *Poverty and Un-British Rule in India, op. cit.*, p. 3.
7. *Ibid.*, p. 24.
8. *Ibid.*, p. 25.

9. *Ibid*., p. 22.
10. *Ibid*., p. 29.
11. *Ibid*.
12. *Ibid*., p. 1.
13. Naoroji, *Dadabhai, Speeches and Writings of Dadabhai Naoroji* (Madras: G.A. Natesan & Co.), n.d., Appendix A, p. 76.
14. Masani, R.P., *Dadabhai Naoroji: The Grand Old Man of India* (London: George Allen and Unwin Ltd.), 1939, p. 453.
15. Naoroji, Dadabhai, *Speeches and Writings of Dadabhai Naoroji, op. cit.*, p. 236.
16. Naoroji, Dadabhai, *Poverty and Un British Rule in India,* op. cit., pp. 187, 236.
17. Ganguli, B.N., *Dadabhai Naoroji and the Drain Theory* (New York: Asia Publishing House), 1965, pp. 92ff.
18. Naoroji, Dadabhai, *Poverty and Un-British Rule in India, op. cit*., p. 30.
19. Black, R.D., *Collision, Economic Thought and the Irish Question, 1817-1870* (Cambridge: Cambridge University Press), 1960, pp. 81-82.
20. Naoroji, Dadabhai, *Poverty and Un-British Rule in India, op. cit*., pp. 43-44.
21. Knight, Robert, *India: A Review of England's Financial Relations Therewith* (London: W. J. Johnson), 1968, p. 430.
22. Viner, Jacob, *Studies in the Theory of International Trade* (London: George Allen and Unwin), 1960, pp. 320-321.
23. Mill, John Stuart, *Principles of Political Economy*, edited by W. J. Ashley (London: Longmans, Green and Co.), 1909, p. 627.
24. Samuelson, Paul, "The Transfer Problem and Transport Costs: The Terms of Trade When Impediments are Absent", *Economic Journal*, Vol. LXII, No. 246 (June), 1952, pp. 278-79.
25. Black, R.D., *Collision, Economic Thought and the Irish Question, 1817-1870, op. cit*., p. 81.
26. *Ibid*., p. 140.
27. Naoroji, Dadabhai, *Poverty and Un-British Rule in India, op. cit*., p. 298.
28. *Ibid*., p. 291.
29. Mill, John Stuart, *Principles of Political Economy*, edited by W.J. Ashley, *op. cit*., 1909, p. 79.
30. Naoroji, Dadabhai, *Poverty and Un-British Rule in India, op. cit*., p. 50.
31. *Ibid*., p. 296.
32. *Ibid*., pp. 48, 170-72.
33. *Ibid*., p. 200.
34. Naoroji, Dadabhai, *The Financial Administration of India* (London: W.J. Johnson), 1871, pp. 4ff.
35. Naoroji, Dadabhai, *Poverty and Un-British Rule in India, op. cit*., p. 181.
36. Naoroji, Dadabhai, *India's Wrongs and Remedy* (Madras: The Commonwealth Office), 1917, p. 13.
37. Naoroji, Dadabhai, *Poverty and Un-British Rule in India, op. cit*., p. 273.

38. *Ibid.*, p. 180.
39. *Ibid.*, p. 272.
40. *Ibid.*, p. 308.
41. *Ibid.*, pp. 321-322.
42. *Ibid.*, p. 322.
43. *Ibid.*, p. 53.
44. Masani, R.P., *Dadabhai Naoroji: The Grand Old Man of India, op. cit.*, 1939, p. 74.
45. Naoroji, Dadabhai, *Poverty and Un-British Rule in India, op. cit.*, p. 190.
46. Naoroji, Dadabhai, *Speeches and Writings of Dadabhai Naoroji, op. cit.*, p. 95.
47. Naoroji, Dadabhai, *Poverty and Un-British Rule in India, op. cit.*, pp. 464-65.
48. Ambirajan, S., *Political Economy and Monetary Management in India, 1766-1914* (Madras: Affiliated East West Press), 1984, pp. 135-6, 145-6.
49. Lipton, Michael and Ravallion, Martin, "Poverty and Policy", Policy Research Working Papers, No. WPS 1130 (Washington: The World Bank), 1993, p. 11.
50. Robinson, Joan, *An Essay on Marxian Economics* (London: Macmillan), 1966, p. 21.
51. Coase, R.H., *Essays on Economics and Economists* (Chicago: The University of Chicago Press), 1994, pp. 15, 33.

4

CONSTITUTIONAL AGITATION

The early Indian leaders had great faith in the efficacy of peaceful and constitutional agitation. Its chief characteristic was that it was linked with the sense of justice and the democratic instincts of the British people. Therefore, the entire strategy of the method of agitation consisted of direct appeal to the British people and the Parliament. The 'moderate' congressmen believed that the British nation was essentially just and good, and that if it could be acquainted with the true state of Indian affairs, all their grievances would be redressed. With this object the Congress organised an effective strategy of agitational activity in England under the dynamic leadership of Dadabhai Naoroji who nursed the British Congress Committee with the support of his Indian British friends with great care.

He had every hope of the success of the Congress work in England, for he believed that India's political aspirations were just and legitimate, and were based on the British promises given to them from time-to-time. In a parliamentary and constitutional manner, Dadabhai wanted to bring the Indian opinion to the notice of the rulers, both in India and England. He assured the authorities that the Congress was 'not a nursery for sedition and rebellion against the British government'. He had never dreamt of severing India's connection with Great Britain. He was, however, of the opinion: "In dealing with Englishmen, make up your minds deliberately, speak clearly, and work perseveringly. Then and then alone can you hope to be listened to."[1] Dadabhai acted as India's representative in England with this conviction.

During the period of his membership of British Parliament (1892-95), and of Welby Commission (1896-1900), Dadabhai kept himself in constant touch with the work and activities of Indian National Congress, with the activities of the British Congress Committee, and its official press organ India, and also with the Indian Parliamentary Committee. These agencies were to promote the Congress movement in England in order to win the British sympathy with support of the Indian demands during the British imperial rule. Dadabhai continued to organise an elaborate and vigorous constitutional agitation in collaboration with his Indian and British friends. After his defeat in parliamentary election in 1895, he was now busy in convincing the British authorities for redressing of Indian grievances.

A staunch constitutionalist, Dadabhai sought to enlighten the British opinion on Indian affairs, and attempted to educate and familiarise the British with the ideals of the Indian National Congress and its demands. He, therefore, set himself to discharge his patriotic duty through personal contacts with the prominent representatives of the British Parliament, British officials and other personalities. He maintained regular correspondence with the British authorities and statesmen. An agitational approach was also tried by him through interviews with the members of Parliament, editors of prominent papers, and other public figures of England. Through correspondence and interviews he conveyed the feeling with regard to the Indian problems, and the Congress attitude towards them. They covered a wide range of subjects connected with the Indian issues, viz. reform of the councils, civil service question, and other matters affecting India and her people.

Interviews and Correspondence

During the last days of Dadabhai's stay in England there was a more intensive campaign for demands in India. British rule on British principles had been undertaken throughout the United Kingdoms than before. His letters and interviews during this period reveal the spirit of his firm determination to continue the fight. Before entering Parliament a representative of the *Daily Graphic* interviewed

Dadabhai on August 8. He put a number of questions to him ranging from his personal life to the object of his political campaign on behalf of India. During the course of his chat with the correspondent, he said: "I stayed in England solely to carry out my political aspirations. I have already held that it would be of the greatest advantage to India; and voice in the Imperial Parliament. What I should most like to see would be the direct representation of the Indian people at Westminster." To a question put to him as to why he wanted to be a member of Parliament, he replied: "But as that privilege has not yet been offered to us, the next best thing we can do is to invite English constituencies to send natives of India to Parliament, when they are otherwise qualified for the trust. That was my reason for coming forward in Finsbury, for my long residence in England has made me as familiar with English political questions as with the wants of my own country."

Asked about the question which Dadabhai considered important to deal in the Parliament, if elected, he stressed on two issues: 'The Indian Civil Service' and the 'Election System for India'. Making his views clear on civil service issue, he said: "...It is less ambitious than many people imagine... I hope to be useful by occasionally calling the attention of the government and of the country to administrative points as they crop up. Anglo-Indian officials do their work on the whole admirably but the best of administrators is not heard by a little parliamentary publicity. Generally speaking, however, these detailed questions are beyond the functions of the British Parliament. They must be fought out in India itself, where we have our Legislative Councils and our own press. The really important work of the House of Commons is to introduce constitutional changes that can only be effected by an Act of Parliament."

Elaborating his views further, he remarked : "The English officials ... are excellent. I must call them angels. The Government of India is solid gold. But you can even buy gold too dear and the price that India pays for English administration is ruinous. No Englishman makes India his home. His children are sent back to England, and a large part of his salary follows to pay for their education. Finally, he returns himself to live in England on a pension provided out of Indian taxes. It is a huge system of absenteeism, like that Irish landlord system. We cannot get rid of it altogether, because we cannot afford to do without some English officials. But we can safely mitigate the evil. And the remedy I propose is to have the examinations for the upper ranks of the civil service held simultaneously in India and in England, so that natives of India may have a better chance of getting appointment."

As regards the election system for India, Dadabhai made his point clear. He went on to state that the existing norms were of nomination and he was in favour of the principle of popular election. "What I propose", he said, "is that a few members of each legislative council should be elected by the same electorate which now choose Municipal Councils." However, Dadabhai did not want the elected members to outvote the government, nor did he imagine the Indian legislature to be the like of British Parliament. He simply wanted the Indian elected members to have the power of advice to the government.[2]

In October 1895, a correspondent of "Great Thoughts" approached Dadabhai, who had laboured with constant assiduity for the advancement of Indian interests in Parliament. He put a question to Dadabhai as to how far his efforts had been successful in throwing open the Indian Civil Service to natives? Dadabhai referred to the June 3, 1893 resolution and claimed it to be an achievement of his efforts. He asserted. "...Now it was the duty of the Government to give effect to that resolution." He voiced his concern about the British unwillingness in acceding to Indian demand as the resolution had met with opposition in the official circles. Pointing out the British policy of apathy and callousness to the Indian affairs, Dadabhai said in despair: "...Therefore, instead of carrying it out, the English government asked the opinion of the Indian legislature, who were well known to be always averse to it. With regard to English affairs, if a resolution is passed relating to any department of the public service, the government is bound to give effect to it. They would never think of first consulting that particular department. But unfortunately, in Indian matter the Indian authorities invariably ignore or resist any resolution or Act which tends to benefit the native population. They do all they can do render it a dead letter. This has been their policy over since 1833, when the first Act was passed 'to treat Englishmen and Indians in His Majesty's Service on the same footing, without regard to race, colour or creed."

Asked to comment on the sincerity of English Government regarding the welfare of India, Dadabhai expressed his doubts over the sincerity of its efforts in improving the lot of Indian people in the existing set up of government machinery. He remarked: "The government may be anxious to promote Indians welfare but unfortunately its action is simply a reflex of the India Office. We are suffering severely from the East India Company to the Crown. We are worse off now than before, because in the days of the old East India Company, we could appeal to Parliament against any unjust act of the Company. But now the government feels that it ought to uphold the Secretary of State for India, whoever he may be, and he is usually a man without any adequate knowledge of Indian affairs, who plays into the hands of the India Council." He then warned "... that this constant violation of pledges, this persistent opposition to Indian interests and the deterioration and impoverishment of the country by an evil administration, must lead, sooner or later, to a rebellion ... I say that forces are gathering vast and powerful, which will succeed in shattering the British Indian Empire, if the present evil and destructive system of administration be not thoroughly reformed ...." In reply to a question what then did he consider would be the wise policy to pursue with regard to India, he replied: "To satisfy her with the justice of British rule, to be honest and honourable in "All our solemn pledges and thus to make Indians feel that to fight for the British rule is to fight for their own hearts and homes and prosperity..."[3]

In another press interview given to G.W. Tooley, and published in *The Humanitarian*, Dadabhai gave expression in emphatic terms to his apprehensions of a revolution in India. He remarked: "Whether your past system was righteous one for India may be matter of argument but to face with an awakened population such as India now has, and will have more and more every year, the present system of government is an absurd and dangerous anachronism. Russia shows what will happen in India unless Indians are given more interest in the government. There will be secret societies."

Adding to his remarks, said he, "...there is immense gratitude in India to England. The present generation feels the benefit it has derived from education .... But succeeding generations will not feel that. They will start where men in India today leave off. They will have no personal memory of old India. However improved the conditions in which they find themselves, they will accept them as a matter of course. They will demand more and if they do not get, they will rebel."[4]

- Again, Dadabhai was interviewed by a representative of "Star" on January 16, 1896. The Grand Old Man represented successfully the opinion and sentiments of Indians, and told the correspondent that Indians would stand firm and loyal to England unless, indeed, some very important mistakes were made by the authorities, he pointed out the many grave evils in the British system of administration, but he admitted the benefits of British connections. However, Indians believed that sooner or later, when the British public came to be conversant with Indian affairs, the evils would sure be remedied. To a question put to Dadabhai that "there is a large amount of fighting power in India", Dadabhai answered: "Quite so, you have in India an amount of fighting power which England can utilize in any part of the world, and which it would be not child's play to resist. You have in the Sikhs, the Rajputs, the Marathas and all the martial races, Hindus or Muhamdans, a warlike breed of some thousands years. Its total strength is not far short of two third of the whole fighting power of Europe ... you have employed them in Africa, China, and in other parts of the world. But England should pay fairly for whatever she takes from India that is to her interest. She should not throw all the burden upon India, as she does in frontier wars likes Chitral, and in many other ways."[5]

Dadabhai then entered into correspondence with the British authorities on subjects of Indian interest. In 1895, the Indian subjects residing then in South Africa contacted Naoroji for his help and assistance against the discriminatory policy of the South African government, which had recognised only the white persons as British subjects. Dadabhai placed the matter before the British Congress Committee, since the issue was treated as important the B.C.C. authorised Dadabhai to take an action. Consequently, he prepared a memorial to present to the Secretary of State for India. He along with W.C. Bannerjee and A.O. Hume waited upon Joseph Chamberlain at the Colonial Office on August 29, 1895 at 1 p.m. The deputation was received by the Secretary of State fairly and sympathetically."[6] As a leader of the deputation, Dadabhai submitted an account of grievances of the British Indians in Africa. Later on he wrote a letter to the Secretary of State reminding him of the deputation and its views on the subject. He said: "You will observe that the whole point of this reference consists in this fact that in ratifying the recent treaty the government of the South African Republic takes upon itself to define 'British subjects', therein referred to as only white persons, thereby depriving all British Indian subjects of the protection for compulsory military service which Her Majesty's Government had found it necessary to insist upon for all British subjects." He concluded the letter with this hope that the Secretary of State would see his own way to disallow this arbitrary limitation of the meaning of the term "British Subjects."[7] Dadabhai's communication, however, did not receive the favour of the Secretary of State for India, who later on expressed his inability to do anything in this regard.[8] While he was a member of the House of Commons, Dadabhai tried to persuade the British authorities, and convince them on the need of indianisation of services. Therefore, in June 1894, he joined the members of Indian Parliamentary Committee, and held an interview with the Secretary of State for India, Sir H.H. Fowle, on the subject. He enquired from him whether something could be done to mitigate the bitter disappointment over not implementation of the June resolution which would be felt in India at the action of Government. But the Secretary of State kept silence, and gave no favourable reply. Nevertheless the question was not abandoned by him, and he pursued the matter with hope.[9] In 1896, in the midst of preliminary work of the Welby Commission, Dadabhai raised the issue of civil service and pressed the authorities in England to give effect to the resolution of June 2, 1893. He initiated correspondence with Lord George Hamilton, the Secretary of State for India; and sent a letter to him on October 12, 1900 along with a cutting from *Kaiser-i-Hind* of September 16, 1900 which expressed strong feelings of Indians against the violation of pledges of equality regarding admission to the services of the state. He wrote: "New generations have received the blessings of education and they now realize the present un-British violation of all pledges, however marked it is by misleading plausible statements.... You will see that there is panic that the authorities were tending towards stopping even in England the competition of Indians for Civil Service, and that the subtle encroachments on the various other services are but a prelude to that disaster.[10] Naoroji sent several such letters to Hamilton conveying the feelings of the educated Indians regarding the civil service question. But all his efforts were of no avail. The conservative leader, then in power, remained unconvinced and unmoved. Hamilton even stopped all communication with the Indian leader on this matter.[11]

Although Hamilton's response was curt and unsympathetic, it could not cool down the enthusiasm of Dadabhai. He raised his voice in another letter in which he said that he would pass over subordinate matters and confine himself to the main point of difference between the two official and non-official views of British rule in India. He asserted: "You announce yourself as a sincere supporter of British rule; you vehemently denounce the conditions and consequences which are inseparable for the maintenance of that rule. This shows that you have not sufficiently known my views. I report that I am a sincere supporter of British rule i.e. of true British rule founded upon, honourable, "faithful and conscientious fulfilment of solemn, pledges." "I denounce the present system, which is not British rule but an un-British, despotic and destructive system which in its declaration of policy is based upon solemn pledges, but which, in its actual practice, is conducted upon the violation of those pledges. The conditions which you lay down as necessary for the maintenance of British rule in India are exactly the conditions which, in their consequences, will destroy instead of maintaining British rule in India."[12] After this, the correspondence between Dadabhai and Hamilton ended virtually, and no reply came from the Secretary of State.

In February 1897, the Diamond Jubilee of Queen Victoria was celebrated throughout the country. Dadabhai reminded the British monarch of her pledges and forced Her Majesty's attention on the sad state of affairs in India during her rule. He urged upon her to modify the British attitude in favour of' India. In a note of February 5, he said : "The most striking consequences of the utter violation of Her Majesty's most sacred pledges and of Acts and Resolutions of Parliament on the extreme impoverishment of the people of British India, and the infliction upon them of all the scourges of the world war, pestilence, and famine, may I hope to look forward to a beneficial change in the present great occasion of Her Majesty's region?"

It is to be noted that Dadabhai maintained correspondence with prominent personalities during this period. But there is little evidence of his exchange of letters between him and the British authorities in larger number. He wrote much to his contemporary friends in India expressing his views and ideas regarding problems of the period. Further, he was discouraged by the response he received from the British in power. Hence little is available of his interviews held, and correspondence conducted with the British authorities.

Agitational Activity through Press

Dadabhai was well acquainted with the power of the press in British politics. So he exercised his great influence through British press for Indian demands and aspirations. He wrote to the press to convey information to support the Indian movement, contradict or correct a wrong information that was sought to be given. His press activity included publication of his articles and letters to India and other British newspapers and journals, circulation of pamphlets and books etc. Its main features were discussion on Indian affairs, information regarding the programme, policies and demands of Indian National Congress, Indian problems, British rule and policies towards India. He made a forceful plea for India and sought solutions of the problems through the columns of British press.

In 1893, he made the civil service question, as his own. He demanded raising of the age for the competition for the covenanted civil service in India, and also holding the competitive examinations simultaneously both in India and in England. He insisted upon the

British Government to give effect to the resolution of Parliament (June 2, 1893). For this issue he used 'the powerful and influential' *The Times* and other journals of England. Of course, the paper published his letter but called it "an exposure of a letter" written by Dadabhai. He, therefore, clarified his stand on the issue and conveyed the feelings of the people of India and the Congress which demanded its implementation. In a letter to the Editor of *Times* dated 9th August 1893 he reported: ". . . That the enthusiasm in India on the passing of the Resolution, is thoroughly genuine and spontaneous, there cannot be the remotest doubt. The subject was taken up by the Indian press, and by the people generally immediately on the arrival of the news by telegram... It is a subject in regard to which the people of India have been crying out for a generation. The Indian National Congress has been putting forward a Resolution demanding this concession and justice....it will be, I think, admitted by everyone that there is a strong feeling really existing." "India is following the gold lessons", he added, "which England has taught her to agitate Constitutionally and persistently for any just cause, and that we are determined to contribute to do."[13]

In another letter addressed to the Editor, *The Daily News*, he again complained of just policy and attitude of the British Parliament and Government by excluding Indians from admission to the higher services. That was a "Clear and open violation of acts of parliament and the most solemn proclamations and promises of the sovereign and British people." Therefore, he appealed to the British people "to remove this blot of injustice and degradation from the British names and to accord to the Indians with British honour and generosity, their true position as British citizens, which is always declared to be their right, by acts of Parliament and Proclamation of the Sovereign."[14]

Dadabhai evoked a keen interest in the British people by his famous book *Poverty and Un-British Rule in India*, in which he brought out the theory of the 'drain' of wealth. He presented facts and statistics to prove the systematic financial burdens thrust upon the shoulders of India. In his book, Dadabhai raised the problem of Indian poverty and of the drain and their intimate connection. He blamed the drain for the misfortunes of India: "Owing to this one unnatural policy of British rule of ignoring India's interests, and making it the drudge for the benefit of England, the whole rule moves in a wrong, unnatural and suicidal groove."[15] He dwelt upon the costliness of existing system of administration which entailed an annual drain of 30 millions sterling without any corresponding return.[16]

Dadabhai's endeavour to place India's condition and demands before the British authorities and press received 'some' sympathies and support. However, his efforts failed in arousing the British conscience to improve India's lot and economic condition as the British policies towards India remained un-effected. But Dadabhai continued to be courageous and outspoken critic of the British rule in India.

Promotion of the Indian Cause from Platform

Dadabhai adopted platform as an important medium to propagate the Indian demands. As one of the early British liberals, laying emphasis on platform activity, asserted that, if anything could move the British electorate on the subject of India, it would be speeches and personal influence of Indian leaders." Therefore, platform activity figured prominently in the constitutional agitation of Dadabhai. He delivered several lectures covering a wide range of subjects pertaining to India. 'Condition of India', 'India', 'India and her wealth', 'British rule in India', were some of the important speeches. These speeches were addressed under the auspices of several associations and organisations of England. Prominent among those were 'London Indian Society', 'Radical Club and Institute', 'Reform Club', 'Liberal Club' etc. The places of his addresses included South Lambeth, Sunderland, Edinburgh, Yorkshire, Liverpool, Manchester and London. In all those meetings he appealed to the public and the British Government to give India 'honour and justice'. He drew the attention of British people towards the extreme poverty in India and demanded the reforms in the administration and policy of righteouness.

As early as in October 1886, he addressed a meeting on the subject entitled 'The political condition of India'. In his address he explained the nature and character of the Indian National Congress, and the advantages of the British rule in enlightening the Indians through western education. At the same time he conveyed the views of educated Indians and their hopes and aspirations for future. Dadabhai saw the political problem in context of the economic problems and their solution. He, therefore, appealed to them "to do their utmost to secure to the natives of India a fulfilment of the solemn promises ... in the confident belief that thereby they would be not simply gratifying the aspirations of educated Indians, but conducting in a marked degree to the prosperity of our great dependency and to the augmentation of our trade with that country". Dadabhai pleaded for fair play in allowing the educated youths of India to compete on equal terms with Englishmen for the offices at the disposal of the Government of India. He said, "India is poor, it is impossible to tax her more, and so she will inevitably remain until her own people have a share in the task of government." "Dadabhai then wished for reform of the Indian Legislative Councils and advocated for the voice of the natives in them. The constitution of those councils was such that the natives had no voice in whatsoever was done for them, in the laws passed for them, in the taxation imposed upon them, or in the manner in which that taxation was spent. India after being for nearly half a century had a sufficient number of people who were prepared to understand that taxation must go without representation. . ."[18]

In another speech he appealed to the British people to see that the 'solemn pledges' of the British are fulfilled in the spirit in which they were made, only then the condition of India would be better and brighter. Dadabhai pointed out the hypocricy of the British when they claimed to have helped every struggling nationality to have self-government to be free to enjoy political rights which God had bestowed upon them. He remarked: "The economic condition of India was utterly unnatural, and at the bottom of all it was the simple fact that foreign rule was a curse and that our rule in India ought to approximate to native rule."[19]

In October 1891, among the many meetings addressed by Dadabhai, there was an important meeting where he identified the British and Indian interests but put India's wants and claims in the most effective manner: "India formed five-sixth of the British Empire, and both must rise or fall together. On the other hand every British question here was of interest to India. Every progress here in the condition of the people would have its reflex action upon the Indian people.... India's poverty or dissatisfaction was a weakness to the BritishEmpire...some fundamental reforms were absolutely needed; and until the Indians themselves were allowed their own representation or voice to reasonable extent both in the Indian Legislative Council and in the Imperial Parliament, and were thus enabled to plead them from their Indian point of view."[20]

Dadabhai dispelled the version of the British that India was made up of two nations, one of Hindus and the other of Muhammadans. He opposed the English views that India could never be a one nation state. The British wanted to divide the Indian people with this assumption, and rule comfortably. He told his British audience in clear terms that the people of India had a common sense to fight for they were just as homogenous as any other nation. He did not agree with the British that India was poor because it was thickly populated. The fact was that the increase in Great Britain during the past ten years was 11.6 per cent against 9.7 per cent in India. The British rule did nothing to prevent famines. He lashed out at the British policy towards India, and narrated its poor economic condition. There had been thirteen famines between 1802 and 1954 in which 5,000,000 people perished, and when railways and irrigation works were supposed to do away with danger of famines during 1860 and 1879, there were sixteen famines in which 12,000,000 people died. Dadabhai, therefore, demanded that "there could be no real prosperity in India until its people were admitted into a share of its government."[21]

In 1895 Dadabhai addressed a conference of Indians at Montagu Mansions Bloomsbury where the subject of reforms of the Indian Legislative Councils was discussed. The conference took up the following resolution for consideration: "That this conference considers the reform and expansion of the supreme and local legislative councils by the admission of considerable proportion elected members and the creation of the similar councils in the Punjab and the central provinces as absolutely essential, and holds that all Budgets should be referred to these councils for adoption." The conference was chaired by Dadabbai, who, in his valedictory address, said: "At present the Indian subjects of the Queen had not the slightest hand in the administration of the affairs of their own country. They, who were born in that country were claiming, if not a predominant at least an equal share in its government. At present they had not a shadow of anything like self-government." He appealed to the people of Great Britain to come to the assistance of their fellow subjects, and help them to obtain some of the blessings of self-government which had done so much for Englishmen.[22]

While Welby Commission was in session and the Indian leaders (as witnesses) were on a visit to England to give evidence, Dadabhai utilised the opportunity to organise a platform campaign throughout the United Kingdom to bring effectively to the notice of the British public the wrongs of India, and the actual condition of India. He was one of the prominent speakers at lectures delivered at South Lambeth, and other places. Public meetings were also held at Yorkshire and Glouchester and the Ealing Division. Though the time was not very much favourable, yet Dadabhai drew large gatherings.

Dadabhai started his platform campaign with a public meeting held on April 26, 1897 in the Library Institute Edinburg. The subject of his address was 'The Coming Collapse of British Rule in India'. He spoke of famine, plague and of the war that had done evil, and made Indian condition more miserable. Proceeding to give a short sketch of the underlying principles of British rule in India, he said that the spirit of the whole administration was pure selfishness, carried out in the most rapacious manner possible. After stating that there were 300 millions of people of India, and arguing that the trade with India might, if 'the country were righteously governed', be so increased that the word 'unemployed' might be erased from British dictionaries. He stated that if the British "continued to bleed a person or a nation, that person or country must in time die. If, however, there was a peaceful revolution of the principles upon which the Indian Government was at conducted, Britain and India would both be blessed."[23]

Speaking at an annual dinner of the London India Society on November 1, 1897, where be was given an enthusiastic reception,

Dadabhai said that whatever India's condition was, that of sorry or joy, Her Majesty always had a share in it. 'What Indians wanted', he told his audience 'was British rule on British principles'. The Queen in her gracious proclamation had given them (Indians) great gifts, upon which they would always take their stand to demand their rights as British citizens. He felt that the Queen had special claims to the attachment and the gratitude of her Indian subjects.[24]

The resolution which was passed at the end of the campaign at a meeting under the auspices of the London India Society on December 28, 1897, is a measure of growth of the feeling of despair and frustration which had over taken Dadabhai after forty years of vain efforts to move the rulers from their unrighteousness and un-British ways. The resolution read as under:

"That of all the evils and terrible misery that India has been suffering from a century and a half, and of which the latest developments are the most deplorable, famine and plague, arising from ever increasing poverty, the stupid and suicidal frontier war and its savagery of the wholesale destruction of villages, unworthy of any people, but far more so of English civilisation, the unwise and suicidal prosecutions for sedition, the absurd and ignorant cry of the disloyalty of the educated Indians and the curtailment of the liberty of the Indian press, the despotism and the general insufficiency of the Administration; of all these and many other minor evils, the main cause is the unrighteous and un-British system of government which produces an unceasing and ever increasing bleeding of the country, and which is maintained by a political hypocracy and continuous subterfuges, unworthy of the British honour and name, and entirely in opposition to the wishes of the British nation and sovereign."

"That unless the present unrighteous and un-British system of government is thoroughly reformed into a righteous and truly British system, destruction to Indian and disaster to the British Empire must be the inevitable result."[25]

In 1898, Dadabhai renewed his efforts to convince the British people and authorities through his platform activity. He went campaigning in Lancheshire and Manchester. He paid special attention to these industrial towns as much harm had been done to

India in the interest of British trade. Dadabhai believed that 'much real good' could be done, if the people of these towns realised what the real condition of India was, and what could really promote the cotton trade between England and India. The first lecture of the series was delivered under the auspices of the East Manchester Liberal Association, in Chorlton Town Hall, Manchester on November 14, 1898. "Are you aware," asked Dadabhai, "that the drain of wealth from India acts a 'protection' of Indian industry". It is a well known economic law that forced payments to another country have the economic effects of a tribute which not only takes from the tributary nation the amount paid but also so raises against them the price of all foreign goods. He told: "...I do not speak with indignation or anger, I am speaking the bare truth that it is most important that the British should be informed and should judge of themselves and not be misled by these who have made it their interest to exploit India, as India had been created by God for that single object."[26]

At Lancheshire, he criticised the British economic policies towards India. He stated: "In England they were able to pay 50s per head but for the purpose of revenuc in India they were not able to pay 5s. Even with free trade in India they had the most wretched trade they could possibly have thought of...in India there were 300,000,000 of people ready to trade with the people of this country, if they would only act righteously and carry out the promiscs made to the natives, looking at the condition of Europe they saw that the natives were ready to fly at each other's throats the "bloated armaments" which were continuously increasing."[27] Hence he desired that the administration of India should be according to merit and not according to colour and creed. Conveying the feelings of Indian people, Dadabhai told that they wanted neither fear nor favour, and if the policy of righteousness would follow, India would be in a far different condition to that in which it was found today. He quoted Macaulay who said that the future of India was unknown but the Indian people, when they understood what true citizenship was, would claim their rights as citizens and that if the people were lifted up from degradation to a higher civilisation the glory would be their own. Dadabhai deplored the policy of the British and concluded:

"The Government of India had continued to act on the most un-English and the most wicked principles. The result was that India was reduced to the utmost poverty and was the poorest country."[28]

He did not agree with the British view that the British rule in India had been a great blessing to the country economically. To him, the system was a vicious and a wrong one in itself, and was carried on in a manner which did not reflect credit upon the British name. He therefore, appealed to the man who "abolished" the slave trade, the so called faddists and fanatics, the philanthropists who raised England to that state of civilisation and advancement of which she was proud to look into the condition of things in India.[29]

At another place, he drew the true picture of the Indian rule under British system, and reminded his audience of England's duty to India: "The British people did not know what they owed to India ... England exported three hundred millions worth of goods to the whole world but what did they send to India, hardly eighteen pence worth per head per annum". Dadabhai stated that the British could not do real good to the Indians unless they did not good to themselves as well. He, therefore, submitted that the government dealing righteously with India, England and even Lanchashire would be rich and prosperous.[30]

Another series of meetings in various parts of England was then arranged for Dadabhai for the year 1899. He addressed large meetings. Notable among these were at Holloway, Warrington (Lanchashire), Dewsbury, Yorkshire. In these meetings the theme of the subject was almost the same. At one place he complained of the apathetic attitude of the British people towards India, and remained of their responsibility. He did not contend that English should make a selfish use of her rights, but at the same time, in as much as the Government of India was supported and maintained by England, the government should take care that India did not set up a fiscal system hostile to vast interests in this country and which could not be tolerated in the country ...[31] On another occasion he delivered a lecture on the condition of India and explained to the British audience the real position of India under British domination. At the same time he told the national character and aspirations of the Indian National Congress, and stressed on the harmonious Indo-

British relations. He made his indictment clear: "The British people have desired, and clearly and emphatically laid it down that the Indian people should be treated and governed exactly like themselves, we are asking no more than that we must be under the same form of Government".[32] On April 20, he addressed a gathering where he described the benefits of British rule in India through western education, freedom of press and speech, but at the same time he laid emphasis on implementation of promises and pledges of the British in true spirit: "The British people had an actual desire to govern India in a spirit of justice, honour and duty ... What he had to complain about was that the mandates they had passed were not carried out, and that they did not trouble to see that they were properly administered".[33]

The same year he addressed a well attended but an important meeting under the auspices of Federation of Liberal Club, Halifax district of Yorkshire. The subject, he chose for his talk, was "Can British rule be popular in India? At the very outset of his speech, Dadabhai criticised the British authorities for their Policies towards India, whereby the Indians were given no opportunity to share in the government of their own country. He, therefore, emphatically said "...if the Indian people were governed on British principles, the popularity in India of British rule would be something they could conceive, the educated people of India knew what British rule was, and they knew that the best rule that could happen to India was that true British ideas should be introduced among them, and that they should be raised to the position of British citizenship."[34] Dadabhai, thus, addressed several such meetings at other places also.

During the year 1900 there was a famine in India, the worst experienced during twenty five years. He was again active, and he had a crowded programme of pubic meetings all over England telling the British public the woes of India. He delivered an address at Edmonton Town Hall on April 29, 1900. "There was", he said, "a chronic state of famine in India. Even in years of average prosperity and average crops scores, of millions had to live on starvation diet. They could not expect to draw blood from a man day after day, and yet expect him to remain healthy." He warned that if no reforms were attempted the famines would go on increasing in intensity, the country would be reduced to utter exhaustion, and he could not venture to forecast what would be the end.[35]

In aid of the Indian Famine Relief Fund, he addressed a meeting on July 1, 1900. Deploring the attitude of apathy of the British towards Indian problems, he presented poor economic condition of Indian people under the Imperial rule, and said: "We Indians are governed by you. You manage our expenditure and our taxes in such a way that where we pay a hundred million pounds of taxation, and hundred million never returns to us ... there is a continual bleeding of about twenty million annually... the time has come when India is bleeding to death. You have brought India to this condition by the constant drain upon the wealth of that country." He then protested: "You impose upon us an immense European Military and Civil Service, you draw from us a heavy taxation. But in the disbursement and the disposal of the taxation we have not the slightest voice ... The Indian government are the masters of all our resources, and they may do what they like with them. We have simply to submit and be bled."[36]

On July 21, 1900 Dadabhai addressed the delegates of the Metropolitan Radical Federation at the Plumstead Radical Club. Moving a resolution on the 'Famine of India', he assailed the British policy which was in distinct contravention of Parliamentary pledges, and Queen's Proclamation. He stated that the 'romance' was that British rule was a blessing to India while the reality was that it was destroying India, and they might depend on it that the destruction of India must ultimately be followed by the destruction of Great Britain. He asked the British to change their policy before it was too late. He expressed his apprehensions that the present famine would be followed by another famine next year because the land had become so dry. Things were going from bad to worse and "it behoved the people of Great Britain to arouse themselves and in the interest of humanity and common justice to adopt such a policy in India as would enable the people to develop the enormous wealth of that country and to enjoy the fruits of their own country."[37]

India Journal, a weekly paper of the British Congress Committee published excerpts from the British paper 'Crydon advertisor' dated October 27, 1900 from a speech of Dadabhai.

The paper supported his views on Indian problems: At Crydon Labour Church last Sunday night at Tom Worth Road Hall, a very practical and interesting address was given on "England's Duty in the present Indian Famine and for the future." The lecture said Mr. Dadabhai Naoroji dealt with the amount of India's wealth drawn every year (£ 30,000,000) and also the millions appropriated for maintaining the Indian Empire, which combined were the real causes of the extreme poverty and destitution. It was the duty of British to pay the cost of all famines and decreases, caused by such impoverishment, out of her own pocket.[38]

Thus, Dadabhai delivered many such speeches in different parts of England. Most of these public meetings attracted a sizeable number of British people. He conveyed the Indian message and feelings to the British public in the most constitutional and effective manner. At last Dr Mullick gave a reception at the Hotel Tudor, Oxford Street in order to afford a number of young Indians interested in the national movement an opportunity of meeting and greeting the Grand Old Man, who, for a long period, had been its inspiring spirit. Sir William Wedderburn, Dadabhai's close associate in his crusade, expressed pleasure he felt in assisting to do honour to his friend. He commended for his good and useful work on behalf of Indian people. Ramesh Chandra Dutt, Sir William Digby and several other British friends recalled some of the main lines of the patriotic work which Dadabhai "had pursued with commendable courage and steadfastness and with notable success."

R.C. Dutt said: "He had a noble example to all Indians left them humbly endeavour to follow in his footsteps. Mr. Naoroji for forty years had done his best to inculcate in the minds of the people of Great Britain ... Mr. Naoroji had done his best to show the real secret of the poverty of India." William Digby, a veteran supporter of Dadabhai, stated "Mr. Naoroji had done that for Indian politics which steam power had accomplished for traction purposes. He had opened up entirely new lines of thought and action; he had demonstrated the only way in which India could once more become a great and prosperous nation."

Dadabhai was delighted to have the opportunity of appealing to the youth of India to come forward and to relieve him who had been working all through his life on behalf of India. In undertaking that responsibility, the young people would be assured, enjoy many advantages, which the oldermen did not posses.... "It is the duty of the younger generation now to come forward and go on with the work which we have inaugurated ... We had to work in the face of the darkest ignorance of Indian matters, but now we have accomplished the labours of the pioneers, now that light has penetrated the darkness, we hold that the time has come when the burden would be shifted on to younger shoulders." He exhorted the future hopes of India to expose the real secret of Indian poverty as he had been time after time, dealing with "reasons and excuses given for that poverty, such as land taxation and the decadence of native industries." He advised his young Indian fellows to carry on the unfinished task for Indian struggle for political and economic advancement on constitutional lines. He had still a firm belief in the British sense of justice and fairplay: "A beginning has been made, the British conscience has been aroused, and hence the outlook is far more hopeful than it has ever been before. Go on impressing on the public mind the real reason of the poverty of India, and the great desirability of giving Indian a voice in the government of their own country, and I am quite sure that once the British tax payer is convinced of the justice of your claims, he will be ready and willing to give you redress."[39]

REFERENCES

1. Report of the Indian National Congress, First Session 1885, Address of Dadabhai Naoroji.
2. *Daily Graphic*, August 9, 1892, Interview reproduced in "*India*' August 26, 1892, pp. 214-15.
3. Vide *India*, October 1895, p. 300.
4. Dadabhai's Interview with G.W. June 1895, R.P. Masani; *Dadabhai Naoraji-The Grand Old Man of India* (London, 1939), pp. 358-59.
5. *India*, February 1896, p. 62, Dadabhai Naoroji on the 'Crisis' loyalty in times of stress and strain.
6. Minutes of the British Congress Committee June 11, August 29 and September 2, 1895 (Vol. Jan. 2, 1896); See *India* October 1895, p. 301, Vol. VI.

7. Dadabbai's letter dated 19th December 1895 addressed to J. Chamberlain, vide Minutes of the B C.C. Vol. Jan. 1894, December 1896.
8. Masani R.P., Dadabhai Naoroji: *The Grand Old Man of India, op. cit.*, p. 471.
9. Kaushik H.P., Indian National Congress in England (A Ph. D. dissertation submitted to the Agra University in 1971), Delhi 1973, p. 76.
10. Quoted in R.P. Masani's *Dadabhai Naoroji: The Grand Old Man of India, op. cit.*, p. 458.
11. Hamilton to Curzon about Dadabhai, 13 December 1900 (Enclosure to the Idler Hamilton to Naoroji) Micro film Reel No. 1613 c. 126/ 1-2. Hamilton Collection.
12. Quoted in R.P. Masani's *Dadabhai Naoroji: The Grand Old Man of India, op. cit.*, p. 460.
13. Dadabhai to the Editor, *Times* 9th August, 1893, Naoroji Papers, N-1, 5d.
14. Naoroji to the Editor, *The Daily News*, 30th November 1898, Naoroji Papers Vol. 1, Letter No. 101.
15. *Dadabhai Naoroji: Poverty and Un-British Rule in India* (New Delhi: Publications Division, Government of India), 1962, p. 211.
16. *Ibid.*, p.211.
17. Alfred Webb to Gokhale, Oct. 23, 1905, Gokhale papers.
18. *Dadabhai Naoroji: Essays, Speeches, Address and Writings* (Bombay: Caxton Printing Works), 1899, pp. 371-72, Dadabhai's address before the members of the Reform Club, London, October 1886.
19. *Ibid.*, p.375. His speech before the members of the North Islington Liberal Club, London, in Nov. 1886.
20. Vide *India,* October 23, 1891, p. 281: Dadabhai Speech at the Nation Liberal Federation at New Castle.
21. Vide *India,* May 1894, Dadabhai's lecture at the Democratic Club London on "Some fallacies about India."
22. *India* January, 1896, p. 16.
23. *India* June 1897, p. 189.
24. *Ibid.*, December 1897, p., 367.
25. Masani, R.P., *Dadabhai Naoroji: The Grand Old Man of India, op. cit.*, p. 396.
26. *India*, November 18, 1898, p. 273.
27. *Ibid.*, November 25, 1898, p. 282.
28. *India*, December 2, 1898, p. 297: His speech at Chorley, Lanchashire on November 16, 1898.
29. *Ibid.*, His address at Manchester on November 22, 1898 under the auspices of Manchester Society for the Protection of Native Races.
30. His address at Achorton, Vide *"India"*, December 9, 1898, p. 311.
31. *India,* February 10, 1999, p. 71: Dadabhai's lecture at North London Liberal Cub and Radical Association on "England's Responsibility for India".
32. His speech at Cooperative Hall, Warrington *India,* February 24, 1899, p. 101 (under the auspices of Liberal Association).

33. *India* April 28, 1899, pp. 212 13: His Speech at Dewsbury.
34. Vide *India*, October 13, 1899, pp. 185-86.
35. Vide *India*, May 4, 1900, pp. 216-17.
36. *Ibid.*, July 6, 1900, pp. 8-9, Dadabhai's Speech at United Methodist Free Church, Waltharnstow, India Must be Bled.
37. *India*, July 27, 1900, p. 50.
38. *Ibid.*, November 2, 1900, p. 225.
39. Vide *India*, October 19, 1900, pp. 201-202.

5

CONCEPT OF SWARAJ

The second most deadly weapon in the world is the hydrogen bomb; the first is the human tongue. The wounds inflicted by it are often worse than those of the second. This fact was unfortunately overlooked by so sagacious a statesman as Lord Curzon at the time of the century. A new life, was then stirring among the people of India. Recalling her glorious past and swayed by a very strong sense of nationalism, the educated classes were looking forward to seeing their motherland regain the place she occupied in ancient times amongst the nations of the world. At that psychological moment Curzon unwittingly wounded their feelings by some of his observations in his University Convocation address: “We have hardly learnt to light the lamp of the soul,” he said. “We have to save the rising generation from walking in false paths and to guide them into right ones.”[1] He was addressing the audience not as the Head of the Government but as one of them, in his capacity as Chancellor of the University of Calcutta for the year 1904. It is not unusual for Chancellors of Universities to stress the importance of character, as he did, but at that tense moment the speech incensed the people who rose as one man to protest against what they regarded as a libel. The critics of Government took the opportunity they wanted for the condemnation of a viceroyalty distinguished, during the earlier years at least, for remarkable sympathy, justice, sagacity and statesmanship.

As the father of Indian Nationalism, Dadabhai rejoiced to find the national spirit of his countrymen that stirred to its depths. The seeds he had scattered broadcast were thus bearing fruit. A new

India was rising. Their complaint, however, he said, was not against individuals. They were fighting against an unjust system. He exhorted the youth of the country to go to the root of the matter and to demand immediate improvement in the system of administration. "Misrule in India" was the subject of his lecture on March 18, 1904, before the members of the Penge and Beckenham Co-operative Society, and the demand for self-government followed. At a gathering of the London Indian Society, responding to the toast of "The President", he declared: "There is only one remedy to the present dishonourable, hypocritical and destructive system. That remedy is self-government under British paramountcy."[2]

Dadabhai then gave a shock to his friends and supporters and created a stir in India by attending an international gathering of Socialists at Amsterdam from August 20, 1904. After having addressed persons belonging to all classes of society in England, he had come to believe that it was the working class that could help India considerably, and that "the people of India should win the daily growing labour force to their side". Apart from that consideration, his innate sympathy with labour had attracted him to the working classes and led him to identify himself with movements to ensure social justice. Testimony of such sympathy comes, in a letter to Mr. R. Masani (February 19, 1929) from George Lansbury:

"He was foremost in his championship of Home Rule for Ireland and for the great measures of social reform which at that time the Radical party was putting forward. I am quite sure that had he been younger and was now living in this country, he would have been one of the foremost men in the Labour Party. I hope one of these days we shall find another such as he, standing for Parliament and standing on the broad basis of international Socialism."[3]

A striking proof of his Socialistic tendencies was given by Dadabhai when at a Conference of Democrats, held on July 29, 1905, under the auspices of the Metropolitan Radical Federation and the National Democratic League at Holborn Town Hall, he moved, as Vice-President of the League, a resolution demanding, the establishment of a universal system of old age pensions, based entirely upon citizens, rights and free from the taint of pauperism. He was, besides, one of the few politicians in England who had mastered the subject of British industrialism. Under the title of The Rights of Labour he had formulated and published a scheme for the establishment of Industrial Commissioners' Courts and for the recognition and protection of labour as a property. If realized, it would have ensured justice to all wage earners and perpetual industrial peace.

At the International Socialist Congress, Dadabhai was received as an honoured guest, representing the people of British India. At one of the sittings the question of India was introduced by a British delegate. The resolution condemning the system of British rule in India, moved by him, hold the familiar ring of the numerous resolutions passed at meetings held under Dadabhai's lead in the United Kingdom. It called upon the workers of Great Britain to enforce upon their Government the abandonment of the "dishonourable" system of administration and the withdrawal of British control.[4]

At that stage the President, Van Kol, a member of the Dutch Parliament, introduced Dadabhai to the assembly. While calling upon him to address the meeting, he asked the audience to rise and stand in silence to mark their respect and commiseration for the suffering millions of India. His words were the signal for one of the most inspiring manifestations of fellow-feeling.[5] As Dadabhai walked slowly to the centre of the platform, he had a rousing reception. The delegates, about a thousand, leapt to their feet and stood uncovered before him in solemn silence. Then their thoughts turned from the Indian people to the dignified presence of their representative himself, and they gave him a tremendous ovation. It was a most impressive spectacle the exploited and downtrodden working men and women of Europe, expressing their goodwill and extending their hand of fellowship to the suffering millions of India.

So overpowered was Dadabhai by feelings of gratitude that when he spoke his voice trembled a little, but as he proceeded to lay the case for India before the delegates, it rang clear and resonant from one end of the concert hall to the other. Most of them did not, before translation, understand a word of what he said, but his delicate features, his refined face, his white hair, and his dignified figure, impressed them immensely. The delegates who understood English punctuated his speech with applause, but when the speech was translated into German and French, the Continental delegates were even more emphatic in their demonstrations of agreement and approval of the downright and, incisive presentation of the case of the afflicted millions on whose behalf he was speaking. They had no idea that an Indian could be so cultured and so captivating as the venerable man who stood before them.

The resolution condemning the system, of British rule in India was carried without opposition. "That means," observed the President, "that this Congress brands Great Britain with the mark of shame for the treatment of India."[6]

Reports of the proceedings of the Socialist Congress were read by people all over the world. Dadabhai had the satisfaction that his appeal on behalf of his motherland attracted considerable attention in the United Kingdom. On the other hand, his association with the Socialists and his impeachment of the British rule roused the ire of the Anglo Indian press in India. Dadabhai was, no doubt, being driven gradually to extremism, at least in his public utterances. Year after year his criticism of the defects of British rule in India increased in acerbity, the result of dire disappointment.

Lord Curzon became more and more unpopular. The division of the historic province of Bengal into two separate provinces with separate capitals antagonized the population. A protest meeting of Indian residents in the United Kingdom was held in London. Dadabhai presided and gave a stirring address, more thrilling and inspiring than any he had ever delivered. "Indians," he said, "have unanimously, earnestly and emphatically declared that the system of rule they are now under should not continue to be."[7]

Asking the audience to consider what that meant, he related what British statesmen had for more than fifty years stated concerning their faulty system of government and recalled the warnings sounded by them from time-to-time to the effect that the continuance of the old system would impel the people to rebel. "More than fifty years ago Mountstuart Elphinstone said: "It is in vain to endeavour to rule them (Indians) on principles only suited to a slavish and ignorant population." And we find not only a continuance of the same old system, but imfacting the people with even more energy and vigour. Now the Indian people have, for the first time, risen and declared that this thing shall not be. Here is a clear issue between the rulers and the people. They have come face to face. The rulers say, "We shall rule, but only as foreign invaders with the result of draining the country of its wealth and killing millions by famines and plague, and starving scores of millions by poverty and destitution, while the ruled are saying for the first time, 'That shall not be'. I regard the day on which the first Calcutta meeting was held as a red-letter day in the annals of India. I am thankful that I have lived to see the birthday of the freedom of the Indian people."[8]

What would be the consequences? Instead of giving his own views, Dadabhai quoted eminent authorities. John Malcolm had prophesied that "the evil" carried with it the seeds of destruction of the Empire. Thomas Munro had said that it would have been more desirable that they should have been expelled from India altogether than that the result of their system of government should be an abasement of a whole people. John Bright had warned his countrymen: "If there be a judgment of nations, as I believe there is as for individuals, our children, in no distant generations, must pay the penalty which we have purchased by neglecting our duty to the populations of India." William Hunter had summed up the gist of one of his lectures: "We should have had an Indian Ireland multiplied fifty-fold on our hands." Lord Hartington had pointed out that the only consequence of exclusion of Indians from the government of their own country could be to make them "desirous of getting rid, in the first instance of their European rulers."[9]

There was no need, continued Dadabhai, for him to refute what Lord Curzon had said in a spirit of peevishness against the character and of the East. He had performed that task before. He rather referred to give illustrations of the political hypocrisy of British statesmen and administrators and to remind them that they were living in glass houses and should not indulge in the pastime of throwing stones. On one aspect of Lord Curzon's conduct he had, however, something to say.

"He does not seem to realize (said Dadabhai) the responsibility of the position in which he had been placed. He is there representing the sovereign of the Empire as Viceroy or second King, the head of a great people, three hundred millions in number, who had possessed civilization for thousands of years and at a time when his forefathers were wandering in the forests here .... His duty as Viceroy is to attract as much as possible and to attach the good feeling of the Indian people to the rule of the British sovereign. What does he do? By this act, he deals a deadly blow to British rule and then by a peculiarly ignorant and petulant speech he creates almost a revolution in the whole empire."

Dadabhai was always ready to give everyone his due. Indeed, he was sometimes blamed for overdoing it in case of British people and officials. While condemning Lord Curzon, he recalled that the Viceroy had claims to the gratitude of the people for what he had done during the early period of his administration.

"I will certainly mention one circumstance in his favour and to his credit. He made a very firm stand against any brutal treatment of the Indian people by Europeans and in so doing caused dissatisfaction to his own countrymen. In that he really did a service not only to India but to the whole British empire. That one act of his shall not be forgotten by Indians...."[10]

"The crisis has come (continued Dadabhai), and the people and the rulers are face to face. The people have for 150 years suffered patiently, and strange to say, their patience has been made a taunt as well as viewed as a credit to them. Often I have been taunted with the fact that three hundred millions of Indians allow themselves to be governed like slaves by a handful of people. But the spell is broken. The old days have passed and the Indian of today looks at the whole position in quite a different light. Now India is becoming restless and it is desirable that the Government should at once realize it."[11]

In order that India's restlessness, may be more pronounced, Dadabhai encouraged Bengal to keep up the agitation.

"The responsibility and the opportunity (he said, in his message to his Bengali friends published in *The Bengalee*, January 5, 1906) that has fallen to your lot is to show that Indians have a backbone, the staying power to the last. If we can once establish this reputation, half of our fight for self-government will be fought and won. I don't care—I'm prepared for ultimate failure. Bengal may remain partitioned. The boycott may at last end, but it will be a great gain if we can once establish the character for organized union and self-sacrifice. One more important result I am looking forward to is the rousing up of the masses, and the present is just the kind of struggle which can accomplish this object. If the people are once roused they will always be ready to follow their leaders."[12]

Dadabhai and Gokhale, were entertained by the Indian residents in the United Kingdom on November 10, at dinner. "The Colonies were prospering with self-government," observed Dadabhai, "whereas India was perishing without it." The agitation of 52 years had not, however, altogether failed. After all, he said: "Great events have been happening abroad, Japan and Russia, and I do not think our present statesmen will shut their eyes to the obvious results of these great events. In the House of Commons I was supported by the Irish, the radical and democratic parties. Outside, we have at present very widely the sympathy and support of the Irish and what I may comprehensively call the people's party. But I depend most of all upon what I see today before me the union of all classes, of my fellow countrymen working together with one heart and with one mind for the great common end. If the whole population of India said once for all that they were determined to have self-government, which was theirs by birth right as well as theirs by pledge, their claim would not be and could not be made in vain."[13]

"Dadabhai had spoken," said Gokhale, "with that overmastering force of conviction which comes from a lifelong and strenuous exertion on behalf of his countrymen. He is "the only man who is entitled to speak in the terms in which he has addressed us. It never will be given to another at any rate for some time to come to occupy so large a place in the hearts of the people of India and that for more than half a century."[14]

Little could Gokhale or anyone else have dreamt at the moment that it would not take even two decades before there would arise another heroic son of India to fill that place, to awaken and rouse the people as he desired and to lead his countrymen manfully towards the goal of self-government and that none other than the hero whom he and Dadabhai both admired and encouraged, as the intrepid commander of the forces of *Satyagraha* in South Africa, the spirited young barrister of the day, Mohandas Karamchand Gandhi.

As the tongue was alert and active, so was the pen during those years, hard at work and unwearied. Scarcely a week passed when Dadabhai did not write to friends or critics or the press to criticise the administration or to ventilate a grievance or to correct a wrong impression about the national movement for self-government.

Here are a few extracts from selected communications:

To J. E. Ellain, July 9, 1898—I may say this only here that when you doubt "whether India would be better off under the Government of any other European power" you do not mean that as a justification of England's evil in India. Then also I do not know in what way or sense you use the word "better off". In its economic condition I do not see how India can be worse off. There are some features in connection with Russia which she will make good use of in trying to make the people of India think that some of the bad features of English rule will not exist in Russian rule. Whether Russia or other power must of necessity occupy India when England is made to leave and whether Russia by becoming worse than England will be able to maintain its power in India are all questions which time alone will solve."[15]

To Romesh Dutt, July 5, 1903—"Lord George Hamilton has made a dead set to get Indians out of the higher services .... Things are growing critical. I have every fear that the attempt which Lord Lytton's Government openly made to stop Indians from competing for the Civil Service here, and which Lord Cranbrook nipped in the bud under Sir Erskine Perry's inspiration, will be sooner or later carried out, if the present Conservative Government continue for any length of time. The time is come when an agitation must be begun for self-government under British Paramountcy. The work will be slow, but every effort needs to be concentrated on this purpose. At my age it will not be my lot to take any long part in this great battle and I am, therefore, the more anxious to see that younger hands and hearts set themselves to work.[16]

To Wacha, January 12, 1905—The very discontent and impatience it (the Congress) has evoked against itself as slow and non-progressive among the rising generation are among its best results or fruit. It is its own evolution and progress.... While there is great necessity for informing the people here, there is as much necessity that the true knowledge of their condition should widely spread among the Indian people themselves. The co-ordination of both is necessary to evolve the required revolution whether it would be peaceful or violent. The character of this revolution will depend upon the wisdom or unwisdom of the British Government and action of the British people.[17]

I hope the next Congress will make a strong pronouncement as to the absolute necessity of self-government as the only remedy for all India's wrongs and needs. Congress should make a clear distinction between two aspects of its duties. The one, a complete change of policy as speedily as possible in the more suitable way leading to self-government—this is Congress's main work—and the second, the hatefulness of the vagaries and failures of the existing administration. The most important of the two is the first.... The whole movement of the Congress must be managed to be backed by the masses.[18]

To the Right Reverend Hundry Codman Potter Bishop of New York, May 14, 1900, in a Telegram from New York— You are reported to have said at a meeting held at the Chamber of Commerce that "the famine was in one sense due to good government. Great Britain had stopped tribal warfare and slaughter and, in consequence, population increased." This, Sir, is the usual Anglo-Indian romance. The reality is quite different, namely that the fundamental cause of "the extreme poverty" of Indians, with its natural consequences of famines, plagues, and every kind of misery, is the destructive system of Government of a bleeding foreign domination.[19]

Bishop of New York to Dadabhai, May 28, 1900—I am bound to add that I do not find myself in agreement with the position you maintain with reference to the English occupancy of India. There was no order nor safety for life or property, nor freedom of person in India until England went there. As a native Indian of high rank recently said to me in Madras, "We have all that you Americans fought for life, liberty and the pursuit of happiness; and if British power were to take itself away from India tomorrow, we should have bloodshed, chaos, and, internecine warfare in endless varieties." I confess, for myself, that the people who are saved from these things should, in my judgment, pay the bill for them.[20]

Dadabhai to Bishop of New York, June 10, 1900—Both, you and the Madrasi gentleman do not seem to have studied and considered the other side. I must say a few words of facts with reference to your Statement that we should pay the bill. We should pay for having been forced to pay every farthing (excepting a very few part payments for very shame) for all the wars, and other circumstances from the very beginning of the English connection, for building up and maintaining the British Indian Empire entirely at our own cost and mainly with our blood...with the reward of being reduced to beggary! We should pay for bleeding us and carrying away clean out of the country hundreds of millions and continue to drain incessantly and unceasingly, or we should pay for what the Viceroys and Famine Commissions sanction.... We should pay for impoverishing us to an extent to which, probably no nation has impoverished another! We should pay for all the consequences of such "other improvements" as famine, plagues...and a chronic state of starvation...! We should pay for the security of our property which, in the most ingenious, scientific and unseen way, is taken away from us by the protectors! We should pay for the security of our lives, which are not left worth living, by providing us with starvation, famines, pestilence etc.! We should pay for the full liberty we enjoy to starve and perish...! We should pay for official Europeans bleeding us and non-official Europeans exploiting our land and labour and natural resources....! In short, we should, pay for a destructive and dishonourable system of government violating Acts of Parliament and the most solemn pledges that ever a people gave to another.[21]

To, Surendranath Banerjee—I wish to write to you, and I am writing, the same to Mr. Motilal Ghosh, that I am grieved to note the differences that are going on between the Patrika and the Bengalee. We cannot afford such quarrelling among themselves.[22]

To the Editor of *The Daily News*, April 17, 1905—In a memorandum of a few statements in the Report of the Indian Famine Commission of 1880, which I submitted on January 4, 1881, to Lord

Hartington, then Secretary of State for India, I said: "India does not get a moment to breathe or revive". 'More Europeans, more Europeans' is the eternal cry, and this very report itself of the Commission is not free from it ....When any question of reform arises, the only remedy that suggests itself to British officials' minds, is "Apply more European leeches, apply more European leeches."[23]

Now the Police Commission Report is out, and other Commissions, Resolutions and Reports for "Reforms" (?) (Heaven save us from those Reforms) have been coming out, with the same eternal cry, "More European leeches, more European leeches" and add to that hideous drain of £ 30,000,000 or more, as if India had no people of her own, and if, she had, as if they existed only to be "bled" and "plundered" with all the ghastly consequences of famine and plague-deaths of millions and starvation of scores of millions! When will this curse end!

To Lord George Hamilton, February 26, 1901—Though the beloved Empress has passed away she has left her will—her greatest, her most noble and her most gracious legacy—to the Indian people worthy of herself and of the great British nation over whom she reigned.... Her son and our new Emperor thrice declared before his very first Council, before the world by his Proclamation issued by yourself, and before Parliament in his very first gracious speech —that he will walk in his good mother's steps. Who now is to be executor of that gracious and glorious will? She prayed: "And may the God of all power grant to us and to those in authority under us the strength to carry out these, our wishes for the good of our people." And who are those who are in authority under us? You are the head of those who are under such authority.... Will you rise to the height of this great and glorious responsibility and opportunity? [24]

Hopes blighted by the bureaucracy were once more revived by the message given by King Edward VII, which formed the theme of another letter sent by Dadabhai to *The Tribune* and *The Morning Leader* (February 22, 1906): "Reading yesterday's proceedings in Parliament I may venture to say that there is and there has been for a long time, a subject of far more serious importance, namely, that under British rule in India millions and millions of human beings are perishing by famine and plague. Would British instincts and sense of duty, conscience and humanity ever lead to deal with the true remedy...? By a divine and blessed inspiration the King himself unconsciously speaks the truth and only remedy."[25]

This antithesis between their viewpoints marked a revolutionary change in the mind of the rising generation. It was not a mere emotional conflict nor was it a contest for power. It was essentially an intellectual conflict, a conflict of ideology. Self-government under British paramountcy was the goal of the old school, freedom from foreign control was the ideal of the new. Constitutional agitation was the accepted method to which the elders wished to adhere; the rebels demanded a radical change in that method. To the ruling class and the European community generally this revolutionary attitude caused no little concern. Heretofore they had affected supreme unconcern about the fulminations of a few extremists and other demonstrations of political unrest in the country. They could not now afford to ignore the striking change that had come over the country. The agitation of the Congress in the past though provoking, had been practically harmless as its leaders had amazingly adhered to the policy of loyalty to the Crown and constitutional methods. There was no menace to the British rule in that sort of agitation, but the young rebels seemed bent on subversive activities. Even they, except a few rabid revolutionaries among them, did not advocate violence at that stage. There was, however, such a weapon as boycott to which they thought they could resort—boycott of British goods and, if necessary, of the councils. Without going beyond the law they could bring the authorities to their knees by refusing to associate themselves with the administration and by withholding from Government all voluntary and honorary service; and they defied any one to say that their ideal was not legitimate or that their method was not constitutional.

The day for the election of the President of the Congress for the year 1906 was drawing nearer. Who should be the President was the question on which a battle royal was expected between the extremists and the moderates. Bal Gangadhar Tilak, who had suffered imprisonment in his fight for freedom of expression, was the choice of the extremists. The moderates, however, were riot enamoured of the agitation carried on by the Maharashtrian hero.

A brain wave of two of their leaders, Surendranath Banerjee and Bhupendranath Basu, enabled them to forestall the move. They sent a cable to Dadabhai, inquiring whether he would accept the appointment, if offered. They felt confident that no one, not even the most rabid extremist would oppose the nomination of one who was universally adored as the Grand Old Man of India, having to his credit an unparalleled record of lifelong patriotic service and self-sacrifice, whose word carried unequalled authority and who alone could be expected at that critical juncture in the history of the Congress to succeed in bringing together the two contending parties.

The weather beaten pilot over whose head storms of over four score years had swept, but who was still a young man at eighty one, promptly replied that he would accept the office, although it meant endangering his health and involved the still greater risk of offending one party or the other or both and risking his reputation as the Nestor of Indian politics. Thus outwitted, the Tilakites acquiesced in the election of Dadabhai at the Calcutta session as President of the Congress for the third time. Even from their view point no better choice, could have been made, as of all members of the Congress outside the extremists' camp, he occupied a position nearest to the advanced wing of the national organization. Belonging to neither party, he had never wavered during the few preceding years in speaking out the language of his heart in condemning the policy of the rulers. Invariably his words breathed fire. For instance, speaking at the Newington Reform Club, he said that they merely told that the British introduced security of property, but that it was only in order that they might carry it away with perfect security. As to the security of life, it was said that the oriental despots used to kill thousands and thousands and harass the people. If that was so, said Dadabhai, "the British Government, with great ingenuity and scientific precision, was killing millions by famines and plagues and starving scores of millions.... The Anglo Indians or the British were like clever surgeons who, with the sharpest scalpels, cut to the very heart, and drew every drop of blood without leaving a scar."[26]

Again in a letter to *The Daily News* dated April 21, 1905, he asked an English correspondent: "Suppose by some mischance England came under French or German or some alien despotic government, would he not, as an Englishman, do his utmost to throw off 'the heaviest of all yokes', the yoke of the stranger, even though all Englishmen were full of all the faults which the Anglo Indians, rightly or wrongly, ascribe to the Indians? Will he not as an Englishman at once tell me: 'Corrupt or not corrupt, faults or no faults, a Briton shall never be a slave?' And yet he coolly justifies and assumes the right divine of making other people slaves!"[27]

For more than three years Dadabhai had been proclaiming that self-government was the only remedy for the ills of India. In a message to Gokhale on the eve of the 1905 Congress, he had stated that no palliative of any kind whatever, no mere alteration of and tinkering at the machinery of administration could or would do any good. The extremists had, therefore, every reason to hope that if he had gone so far in the year 1905 as asserting the birth-right of his countrymen, he might in 1906 go a step further.

The keynote of Dadabhai's address at the Calcutta Congress, the largest political gathering witnessed in India until then, was Swaraj. It was the first occasion when the demand for self-government was formulated and put forward from the platform of the Congress. The whole country was thrilled. It was the first session of the national organization after its coming of age. It was time, said the revered Dada of the nation to the audience, to consider what their future should be. Were they British citizens or not? The moment Indians, had come under the British flag, they had become free British citizens and their rights as such were beyond question. They had every reason to claim all such rights as their birth-right and also—as rights solemnly pledged to them. "Just as the administration of the United Kingdom in all services, departments and details was in the hands of the people of that country, so should it be," said he, "in India. As in the United Kingdom and the Colonies all taxation and legislation and all power of spending the taxes were in the hands of the representatives of the people, so should it be in India, and the financial relations between England and India must be adjusted on a footing of equality."[28]

He was not asking for any favour. He was pleading for justice. The whole matter, he said, could be compressed in one word, "self-government", Swaraj like that of the United Kingdom or the Colonies.

The whole machinery of government could not, certainly, be broken up and the system of self-government introduced forthwith. The time had, however, arrived for the transfer of power to begin.

Coming to the crucial question of methods, he proceeded to give a full and frank exposition of the problem as he visualized it: "Since my early efforts, I must say that I have felt so many disappointments as would be sufficient to break any heart and lead one to despair and even, I am afraid, to rebel ... But I have not despaired... You may think it strange. I stand before you with hopefulness. I have not despaired for one reason and I am hopeful for another reason. I have not despaired under the influence of the good English word, which has been the rule of my life. That word is 'Persevere'....As we proceed, we may adopt such means as may be suitable at every stage but persevere we must to the end. Now the reason of my hopefulness after all my disappointments. And this also under the influence of one word, 'Revival'—the present revival of the true old spirit and instinct of liberty and free British institutions in the hearts of the leading statesmen of the day....Within the short life that may be yet vouchsafed to me, I hope to see a loyal, honest, honourable, and conscientious adoption of the policy of self government for India and a beginning made at once towards that end."[29]

Turning to the burning question of the day, partition of Bengal, he characterized it as a "bad blunder for England." He hoped it would be rectified. Swadeshi existed in Bombay for many years. "I am a free trader," he said. "I am a member in the Executive Committee of the Cobden Club for twenty years, and yet I saw that Swadeshi is a forced necessity for India in its unnatural economic muddle. As long as the economic condition remains unnatural and impoverishing ... the talk of applying economic laws to the condition of India is adding insult to injury."[30]

The address ended with a fervent plea for unity: "I do not know what good fortune may be in store for me during the short period that may be left to me, and if I can leave a word of affection and devotion for my country and countrymen, I say: "Be united, persevere and achieve self-government so that the millions now perishing by poverty, famine and plague, and the scores of millions that are starving on scanty subsistence may be saved, and India may once more occupy her proud position of yore among the greatest and civilized nations of the world."[31]

It was meant to be, and was in fact a conciliatory speech. Strongly controversial topics such as Swadeshi were disposed of in a few words. "Boycott" was not even mentioned. It was not quite clear where Dadabhai stood or how far he would allow the advanced wing of the Congress to go. It had not got what it wanted. Instead of recommending any change in methods and tactics, Dadabhai had simply asked his countrymen once more to have faith in the British people's sense of justice. He had held out hopes in the revival of Liberalism in England, whereas India had abandoned hopes of getting redress even at the hands of the Liberal Party. On the other hand, the moderates were pleased. The Bengalee hailed the address as "the political gospel for the new era", while Bande Mataram shed tears on "the great refusal". To the Anglo-Indian community and the press, Swaraj had always been a nightmare. The presidential address was, therefore, considered a surrender to the extremists.

During the meeting of the Subjects Committee there were violent scenes over the refusal to submit resolutions for extending boycott all over India. Some of the extremists left the meeting, headed by Bipan Chandra Pal and Khaparde. There was, for a while, jubilation in the camp of Anglo-Indians hostile to the Congress, but those who welcomed the split were amazed to find a display of unanimity at the final session. There was heated discussion, no doubt, on the resolution moved by Ambica Charan Mazumdar, seconded by Pal that the boycott movement inaugurated in Bengal, by way of protest against the partition of that Province, "was and is legitimate". The extremists put their own interpretation on the resolution, claiming that it applied to the whole country, but the moderates interpreted it as confined to Bengal only.

A united front had, however, to be presented to the world outside. Tilak had the patriotism and the statesmanship to realize it. Despite his disappointments he struck a refreshing note of unity. Supporting the Swadeshi resolution, he said he was glad that they had arrived at a satisfactory solution because their Anglo-Indian friends had predicted that the twenty-second Congress would meet with premature death immediately after the attainment of the age of majority. The prediction, however, had been falsified under the able, impartial and judicious guidance of the veteran leader in the chair. All differences, continued Tilak, had been squared; both parties had approached the question in a spirit of conciliation and met halfway.

REFERENCES

1. Masani, R.P., *Dadabhai Naoroji* (New Delhi: Publications Division, Ministry of Information and Broadcasting, Government of India), 1975, p. 130.
2. *Ibid.*
3. Patwardhan, R.P. (ed.), *Dadabhai Naoroji Correspondence* (Bombay: Allied Publishers), 1977.
4. *Ibid.*
5. Masani, R.P., *Dadabhai Naoroji, op. cit.*, 1975, p. 133.
6. Naoroji, Dadabhai, *Speeches and Writings of Dadabhai Naoroji* (Madras: G. A. Natesan & Co.), n.d.
7. *Ibid.*
8. *Ibid.*
9. *Ibid.*
10. Masani, R.P., *Dadabhai Naoroji, op. cit.*, 1975, p. 136.
11. *Ibid.*
12. *The Bengalee*, 05-01-1906.
13. *Ibid.*
14. *Ibid.*
15. Patwardhan, R.P. (ed.), *Dadabhai Naoroji Correspondence, op. cit.*, 09-07 1898.
16. *Ibid.*, 05-07-1903.
17. *Ibid.*, 12-01-1905.
18. *Ibid.*
19. *Ibid.*, 14-05-1900.
20. *Ibid.*, 28-05-1900.
21. *Ibid.*, 10-06-1900.
22. *Ibid.*
23. *Ibid.*, 17-04-1905
24. *Ibid.*, 26-02-1901.
25. *The Tribune* and *The Morning Leader*, 22-02-1906.
26. Masani, R.P., *Dadabhai Naoroji, op. cit.*, 1975, p. 149.
27. *The Daily News*, 21-04-1905.
28. Naoroji, Dadabhai, *Speeches and Writings of Dadabhai Naoroji, op. cit.*
29. *Ibid.*
30. *Ibid.*
31. *Ibid.*

BIBLIOGRAPHY

Ambirajan, S., *Political Economy and Monetary Management in India*, 1766-1914 (Madras: Affiliated East West Press), 1984.

Asiatic Quarterly Review, April 1893.

Banerji, A.K., *Aspects of Indo British Economic Relations, 1858-1898* (Bombay: Oxford University Press), 1982.

Banerji, Arun, *Finances in the Early Raj: Investments and the External Sector* (New Delhi: Sage Publications), 1995.

Black, R.D., *Collison, Economic Thought and the Irish Question, 1817-1870* (Cambridge: Cambridge University Press),1960.

Blaug, Mark, *"Kuhn versus Lakatos or Paradigms versus Research Programmes in the history of economics"* in *Method and Appraisal in Economics,* edited by Spiro J. Latsis (Cambridge: Cambridge University Press), 1976.

Blaug, Mark, *Economic Theory in Retrospect* (Cambridge: Cambridge University Press), 1978.

Central Statistical Organisation, *National Accounts Statistics: Sources and Methods* (New Delhi: Department of Statistics, Ministry of Planning, Government of India), 1989.

Chakarverty, S., "Drain Theory of Dadabhai Naoroji," a paper presented to the Indian Economic Association Conference, December, 1968.

Chandra, Bipan, *The Rise and Growth of Economic Nationalism in India, Economic Policies of Indian National Leadership 1880-1905* (New Delhi: People's Publishing House), 1966.

Charlesworth, Neil, *British Rule and the Indian Economy*, 1800-1914 (London: Macmillan), 1982.

Chaudhuri, K.N., "India's International Economy in the Nineteenth Century: An Historical Survey", *Modern Asian Studies*, Vol. 2, No.1 (January), 1968.

Coase, R.H., *Essays on Economics and Economists* (Chicago: The University of Chicago Press), 1994.

Clark, Colin, *Conditions of Economic Progress* (London: MacMillan and Co. Ltd.), 1949.

Dantwala, M.L., *Poverty in India, Then and Now, 1870-1970* (Delhi: MacMillan), 1973.

Das, M.N., *Indian National Congress versus The British*, Vol. 1, (1885-1918), Delhi, 1978.

Dasgupta, Ajit K., *A History of Indian Economic Thought* (London: Routledge), 1993.

Datta, Bhabatosh, *Indian Economic Thought—Twentieeth Century Perspectives* (New Delhi: Tata McGraw Hill Publishing Company Ltd.), 1962.

Datta, Bhabatosh, *The Evolution of Economic Thinking in India* (Dr. P.N. Banerjea Memorial Lectures, 1961) (Calcutta: Federation Hall Society), 1962.

Desai, S.S.M., *Economic History of India* (New Delhi: Himalaya Publishing House), 1990.

Dimand, Robert W., "I have no Great Faith in Political Arithmetick: Adam Smith and Quantitative Political Economy" in *Measurement, Quantification and Economic Analysis: Numeracy in Economics* edited by Ingrid H. Rima (London: Routledge), 1995.

Dutt, Romesh, *The Economic History of India Under the British Rule*, Vols. 1, 2 & 3 (New Delhi: Publications Division, Ministry of Information and Broadcasting, Government of India), 1963.

Ganguli, B.N., *Dadabhai Naoroji and the Drain Theory* (New York: Asia Publishing House), 1965.

Ganguli, B.N., *Indian Economic Thought: Nineteenth Century Perspectives* (New Delhi: Tata McGraw Hill Publishing Co. Ltd.), 1978.

Georgescu-Roegen, Nicholas, *Analytical Economics: Issues and Problems* (Cambridge, Massachusetts: Harvard University Press), 1966.

Gopalakrishnan, P.K., *Development of Economic Ideas in India, 1880-1914* (Gravenhange: Institute of Social Studies), 1954.

Gupta, Ajit Das, *A History of Indian Economic Thought* (Madras: T. R. Publications Private Limited), 1990.

Ingrid H. Rima (ed.), *Measurement, Quantification and Economic Analysis: Numeracy in Economics* (London: Routledge), 1995.

Jenk, L.H., "British Experience with Foreign Investments," *Journal of Economic History*, 1944.

Jha, Shiva Chandra, *A History of Indian Economic Thought* (Calcutta: Firma K. L. Mukhopadhyay), 1981.

Karmakar, A.D., *Economic Thought of Modern India* (Dharwar: J.S.S. Institute of Economic Research), 1974.

Karunakaran, K.P., *Indian Politics from Dadabhai Naoroji to Gandhi: A Study of the Political Ideas of Modern India* (New Delhi: Gitanjali Publishing House), 1988.

Khan, Sir Syed Ahmad, *The Complete Collection of Lectures and Speeches* (ed.) by M.I. Gujarati (Lahore: Nawal Kishore), n.d.

Knight, Robert, *India: A Review of England's Financial Relations Therewith* (London: W.J. Johnson), 1968.

Kurdas, Cigdern, "The Whig Historian on Adam Smith: Paul Samuelson's Canonical Model", *History of Economics Society Bulletin*, Vol. X, No. 1 (Spring), 1988.

Lakatos, Imre, "Falsification and the Methodology of Scientific Research Programmes" in *Criticism and the Growth of Knowledge* edited by Imre Lakatos and Alan Musgrave (Cambridge: Cambridge University Press), 1970.

Lipton, Michael and Ravallion, Martin, "Poverty and Policy", *Policy Research Working Papers, No. WPS 1130* (Washington: The World Bank), 1993.

Loganathan, *A History of Economic Thought* (New Delhi: S. Chand and Company Ltd.), 1987.

Madan, G.R., *Economic Thinking in India* (Madras: S. Chand & Co.), 1982

Mahajan, O.P., *The Drain Theory: Fact and Fiction* (IEA), 1968.

Majumdar, Jatindra, Kumar (ed.), *Indian Speeches and Documents on British Rule 1821-1918* (London), 1937.

Masani, R.P., *Builders of Modern India—Dadabhai Naoroji* (New Delhi: Publications Division, Ministry of Information and Broadcasting, Government of India), 1960.

Masani, R.P., *Dadabhai Naoroji: The Grand Old Man of India* (London: George Allen and Unwin Ltd.), 1939.

Masani, R.P., *Dadabhai Naoroji* (New Delhi: Publications Division, Ministry of Information and Broadcasting, Government of India), 1960.

Max Karl and Fredrick Engels, *On Colonialism* (Moscow: Progress Publishers), 1965.

Mill, John Stuart, *Principles of Political Economy*, edited by W.J. Ashley (London: Longmans, Green and Co), 1909.

Morrison Sir, *Theodore Economic Transition in India* (London: London Press), 1916.

Mukherjee, M., *National Income of India, Trends and Structure* (Calcutta: Statistical Publishing Society), 1969.

Mukerjee, Tapan, "Theory of Economic Drain: Impact of British Rule on the Indian Economy: 1840-1900" in *Economic Imperialism: A Book of Readings*, edited by K.E. Boulding and Tapan Mukherjee (Ann Arbor: The University of Michigan Press), 1972.

Nabers, Lawrence, 'The Positive and Genetic Approaches' in *The Structure of Economic Science Essays on Methodology*, edited by Sherman Roy Krupp (Englewood Cliffs, N.J.: Prentice Hall, Inc.), 1966.

Nanda, B.R., *Gokhale, Gandhi and the Nehru's Studies in Indian Nationalsim* (London: George Allen and Unwin Ltd.), 1990.

Naoroji, Dadabhai, *A Sketch of His Life and Career* (Madras: G.A. Natesan & Co.), 1920.

Naoroji, Dadabhai, *Speeches and Writings of Dadabhai Naoroji* (Madras: G.A. Natesan & Co.), n.d.

Naoroji, Dadabhai, *The Financial Administration of India* (London: W.J. Johnson), 1871.

Naoroji, Dadabhai, Essays, Speeches, Addresses and Writings (Bombay: Caxton Printing Works), 1899.

Naoroji, Dadabhai, *Poverty and Un-British Rule in India* (New Delhi: Publications Division, Government of India), 1962.

Naoroji, Dadabhai, *India's Wrongs and Remedy* (Madras: The Commonwealth Office), 1917.

Pandey, Shridhar, *Economic History of India* (Hindi) (Patna: Motilal Banarsidass), 1991.

Parekh, C.L, *Essays, Speeches, Address and Writings (on Indian Politics) of the Hon'ble Dadabhai Naoroji* (Bombay: Asia Publishers), 1887.

Parliamentary Debates (House of Commons) Fourth Series Vol. XI, 13th April 1893.

Patwardhan, R.P. (ed.), *Dadabhai Naoroji Correspondence, Child and Youth Section*, Vol. II, Part I (Bombay: Allied Publishers), 1977.

Paul, R.R., History of Economic Thought (New Delhi: Kalyani Publishers), 1979.

Ranade, Mahadev Govind, *Ranade's Economic Writings edited* by Bipan Chandra (New Delhi: Gian Publishing House), 1990.

Rao, V.K.R.V., *An Essay on India's National Income, 1925-1929* (London: George Allen and Unwin), 1939.

Rashid, Salim, "Does a Famous Economist Deserve Special Standards? A Critical Note on Adam Smith Scholarship," *History of Economics Society Bulletin*, Vol. XI, No. 2 (Fall), 1989.

Robinson, Joan, *An Essay on Marxian Economics* (London: Macmillan), 1966.

Rogin, Leo, *The Meaning and Validity of Economic Theory* (New York: Harper and Bros.), 1956.

Samuelson, Paul, "The Transfer Problem and Transport Costs: The Terms of Trade when Impediments are Absent," *Economic Journal,* Vol. LXII, No. 246 (June), 1952.

Schumpeter, J.A., *History of Economic Analysis* (New York: Oxford University Press), 1954.

Scweinitz, Jr. Karl de, *The Rise and Fall of British India: Imperialism as Inequality* (London: Methuen & Co.), 1983.

Sen, R.K. and Mandal, S.K., *Ranade's Analysis of India's Population Problem and the Views of Naoroji and Doskar*, I.E.A., 75th Conference Volume, 1992.

Sen, Sunanda, *Colonies and the Empire India, 1890-1914* (Hyderabad: Orient Longman), 1922.

Shanmugasundaram, V., *Indian Economic Thought and Policy* (New Delhi: S. Chand and Company Ltd.), 1981.

Singh, V.B., *From Naoroji to Nehru* (New Delhi: McMillan), 1975.

Stark, Werner, *The History of Economics in its Relation to Social Development* (London: Kegan Paul, Trench, Trubner & Co.), 1944.

Studenski, Paul, *The Income of Nations* (New York: New York University Press), 1958.

Viner, Jacob, *Studies in the Theory of International Trade* (London: George Allen and Unwin), 1960.

World Bank, *World Development Report, 1990* (Oxford: Oxford University Press), 1990.

Zaidi, Moin A., *Dadabhai Naoroji, Speeches and Writings* (New Delhi: Indian Institute of Applied Political Research), Vol. 1, 1985.

Index